MW01616423

THE SUPER-CHARGED WORLD OF CHEMISTRY
PARTS 1,2,&3
STUDY SIDEKICK
(1ST EDITION)

WRITTEN BY THE STANDARD DEVIANTS®
ACADEMIC TEAM, INCLUDING:

Dr. David Rowley, Ph.D.

Kristie Wingenbach

Rachel Galvin

EDITED BY:

Rachel Galvin

CONTRIBUTING EDITORS:

Dr. David Ramaker, Ph.D.

Dr. John Schreifels, Ph.D.

ART DIRECTION & GRAPHIC DESIGN BY:

C. Christopher Stevens

ILLUSTRATION & ANIMATION BY:

Sarah W. Fry

Bryan Knight

HOW TO REACH US AT CEREBELLUM:

phone: 800-238-9669

e-mail: cerebellum@mindspring.com

web address: www.cerebellum.com

©1997 CEREBELLUM CORPORATION

2

Other Subjects From Cerebellum

The Eye-Popping World of
FINANCIAL ACCOUNTING 1 & 2

The Deep-Fried World of
ORGANIC CHEMISTRY 1, 2, & 3

The Salsa-riffic World of
SPANISH

The Stimulating World of
PSYCHOLOGY

The High-Stakes World of
STATISTICS

The Really Big World of
ASTRONOMY

The Dissected World of
BIOLOGY

The Dangerous World of
PRE-CALCULUS 1 & 2

The Unbelievable World of
PHYSICS

The Creepy, Crawly World of
CALCULUS 1 & 2

The Zany World of
BASIC MATH

The Rockin' World of
GEOLOGY 1 & 2

The Twisted World of
TRIGONOMETRY 1 & 2

The Wild and Wacky World of
FINANCE 1, 2, & 3

The Adventurous World of
ALGEBRA 1 & 2

For an updated list of titles available, check our web site:

www.cerebellum.com

Printed in the beautiful U.S.A.

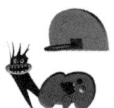

HOW TO USE THIS BOOK

FOREWORD

CHECK OUT THE VIDEOS. Please notice the plural **videos!** This single Study Sidekick corresponds to three of our Video Course Reviews: *The Super-Charged World of Chemistry Parts 1, 2, and 3*. This workbook, when used along with the three videos, will help cut through the chemical fog.

FOLLOW ALONG. The **VIDEO NOTES** section does your work for you! We've already taken all of your notes—all you have to do is follow along with the videos. We've even given you **VIDEO TIME CODES** for all three videos. Just reset your VCR counter to 0:00:00 when the Cerebellum logo appears at the beginning of each tape. These clocks **0:00:00** give you the time code for each important section so you know where to fast-forward to! This will enable you to learn and retain material much more effectively. Just stop the tape after a difficult section and read through your notes!

EASY-TO-FIND INFO. So you'll know which video we're talking about, we've put these markers at the bottom of each page: **V1, V2,** and **V3. V1** means you're in a section that covers material from *The Super-Charged World of Chemistry Part 1*, **V2** means you're in a section that covers *The Super-Charged World of Chemistry Part 2*, and **V3** means you're in a section that covers *The Super-Charged World of Chemistry Part 3*.

4

LEARN NEW STUFF. Unfortunately, we just can't include everything about chemistry in three videos. The **OTHER IMPORTANT STUFF** section gives other cool facts you'll need to ace your tests.

TEST YOURSELF. QUIZZES and **PRACTICE EXAMS** allow you to test yourself and make sure you've covered all the bases. The answers appear at the back––*don't cheat!*

HAVE FUN. The book is chock-full of diversions and stress relievers, and there's a neat flippy picture on the bottom of each page.

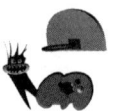

TABLE OF CONTENTS

STUDY SIDEKICK

6

VIDEO NOTES

TABLE OF CONTENTS

The Super-Charged World of Chemistry Parts 1, 2, & 3

7

TABLE OF CONTENTS

OTHER IMPORTANT STUFF

VIDEO TIME CODE

The Super-Charged World of Chemistry Part 1

STUDY SIDEKICK

12

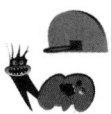

CHEM SAFETY GUY says,

WARNING:

What you are about to read may cause your eyes to cross and your toes to curl and uncurl convulsively. You may feel compelled to pull out your hair or use the bathroom.

BUT DON'T FREAK.

14

WHAT IS CHEMISTRY?

`0:02:02`

Chemistry is more than just a meltdown of elements, electrons, test tubes, and explosive experiments––it's the study of the stuff around us and how it changes. Chemists can tell you what your body is made of, how antifreeze keeps your radiator from cracking in the winter, and why iron rusts. We study chemical reactions in biology, geology, engineering, physics, and the chemical reactions occurring between the beef jerky and corn chips that you had for dinner.

`0:03:24`

SECTION A: INTRODUCTION TO MATTER

Chemists study how energy and matter interact. **Matter** is anything that takes up space and has mass. When you think matter, think *space* and *mass*.

We use the terms "mass" and "weight" interchangeably in this Study Sidekick, but you can't do that all the time. For our purposes, however, they are the same doggone thing––so don't let the two terms confuse you.

VIDEO NOTES

The Super-Charged World of Chemistry Part 1

Matter may take up just a little space with a lot of mass, like a rhinoceros does, or it may take up a whole lot of space with very little mass, like air. Energy isn't matter because it doesn't take up space or have mass. Rather, energy acts on matter: For instance, when Bobo the Clown's rear end is on fire, he sees the light and feels the heat emanating from the flames roasting his hind parts. Both light and heat are examples of energy doing things to matter.

Matter comes in three states: solid, liquid, and gas. When matter changes from one state to another, like when water evaporates or ice melts, we call that a **change of state.**

Each solid, liquid, or gas has a unique set of characteristics (properties) that we can observe and measure. There are two kinds of properties: physical and chemical. **Physical properties** are physical attributes such as temperature, height, weight, consistency, odor, and hardness. You can measure physical properties without changing the identity of the matter you're studying.

Melting point is a physical property. As ice becomes water, it changes its physical state from solid to liquid. Water doesn't look like ice, but its basic identity is the same; so when we measure the melting point of

0:04:44

0:04:58

I was so touched by her that, I don't know, after fifteen minutes I wanted to marry her, and after half an hour I completely gave up the idea of snatching her purse.

– Woody Allen in *Take the Money and Run*

point of ice, we are measuring a physical property. Boiling point, freezing point, and ability to evaporate are all physical properties of matter. (The ability to walk through walls is not a property, however, but a feat of supernatural prowess.)

Chemical properties of matter describe the way matter acts during chemical reactions. For instance, one of water's chemical properties is its ability to decompose into the elements hydrogen and oxygen. Once water decomposes, it isn't water any more.

WARNING: You can't observe chemical properties without changing the matter!

You can only find chemical properties through experimentation. This is one reason why experiments are such an important part of chemistry.

There are two different types of matter: substances and mixtures. A **substance** has the same physical and chemical properties throughout it. Distilled water is a substance and an example of **homogeneous matter**, since every part of it has the same physical and chemical properties.

A mixture is matter that can be separated into two or more substances. A sedimentary rock is a mixture and an example of **heterogeneous matter**, since it is made up of fragments of other rocks. A heterogenous pastry would be made up of recycled fragments of other pastries, like croissants, danishes, and donuts.

A mixture of substances is considered homogeneous if the mixture has the same physical and chemical properties throughout it. A homogeneous mixture is also called a **solution**. Adding some blue dye to a glass of water produces a homogeneous mixture, because once the dye dissolves in the water, the entire mixture has the same chemical and physical properties throughout.

On the other hand, a heterogeneous mixture has different chemical and physical properties that remain separate from each other. Orange juice is a heterogeneous mixture, since it's possible to strain the bits of orange out.

You can separate both homogeneous and heterogeneous mixtures into their component parts by physical means such as distillation and filtration. Substances, however, must be broken down by chemical means such as the addition of heat or an electrical current. When you separate a substance, you get either an **element** or a **compound**.

DEIONIZED INFORMATION:

Here's a good way to think of the difference between physical and chemical properties:

If you change the property and the matter only *looks* different (its identity hasn't changed), you're dealing with a physical property.

If you change the property and the matter's identity has changed, like when water breaks down into the elements hydrogen and oxygen, you're dealing with a chemical property.

Compound: a substance with two or more kinds of atoms combined in fixed proportions.

Element: a substance that cannot be broken down chemically into a simpler substance.

.A substance has the same physical properties throughout.

.Homogeneous mixtures of substances, called solutions, also have the same physical properties throughout.

.Physical properties can be seen without changing the identity of matter, but to see chemical properties, you must change the identity of matter.

.Heterogeneous matter is a mixture of various kinds of matter and has different chemical and physical properties in different parts of the mixture.

.You can separate heterogeneous and homogeneous mixtures into their components by physical means. Substances, however, can only be separated into their components by chemical means.

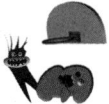

VIDEO NOTES

The Super-Charged World of Chemistry Part 1

SECTION B: THE ELEMENTS

`0:10:24`

Atoms are the building blocks for absolutely everything. Either by themselves or with other atoms, they make up all matter. An element is a substance that contains only one kind of atom. Right now, there are 110 known atoms, and each has unique characteristics that distinguish it from the others.

`0:10:33`

All the elements are listed in the periodic table. They fall into three categories: metals, nonmetals, and metalloids. The metals are on the left side and in the middle of the table, with the exception of hydrogen, which is on the left side although it is not a metal.

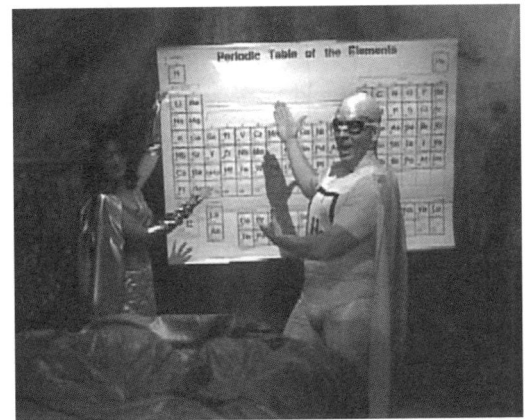

The nonmetals are on the right, and the metalloids are in a narrow diagonal strip between the metals and the nonmetals.

Helium, neon, argon, krypton, xenon, and radon are called the "noble gases" because they do not combine easily with other elements to form compounds. (So that means krypton does not combine easily with carbon to form kryptonite. Superman is safe...for now.)

`0:13:10`

V1

`0:11:46`

`0:12:07`

Compounds are combinations of elements. A compound is a substance that has two or more kinds of atoms combined in fixed proportions. A single unit of a compound is called a molecule. Water is a compound of the elements hydrogen and oxygen. To form a water molecule, two hydrogen atoms and one oxygen atom combine through a chemical reaction. The chemical reaction makes the atoms move from their original combinations to form new combinations.

It's important to remember that during a chemical reaction, the atoms themselves don't change. There are not more or fewer atoms––they just rearrange to form the compound.

A molecule's **molecular formula** shows the exact number of atoms of each element that makes up the molecule. For water, the molecular formula is H_2O. "H_2" means there are 2 hydrogen atoms in the molecule, and since "O" doesn't have any subscript numbers, we know there is only one oxygen atom.

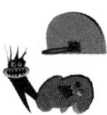

UNITS OF MEASURE

SECTION A: THE METRIC SYSTEM AND SI UNITS

Like physics and all the sciences, chemistry uses the metric system. You are probably used to the English system, but the metric system is even simpler when you're converting from larger to smaller units (or vice versa), since it's just a matter of prefixes. Here's a refresher on how it works.

THIS CHART shows the most commonly used prefixes. You will see these everywhere, so it's useful to familiarize yourself with them, particularly *nano* through *mega*.

METRIC SYSTEM PREFIXES

10^{-15}	femto-	one quadrillionth
10^{-12}	pico-	one trillionth
10^{-9}	nano-	one billionth
10^{-6}	micro-	one millionth
10^{-3}	milli-	one thousandth
10^{-2}	centi-	one hundredth
10^{3}	kilo-	one thousand times
10^{6}	mega-	one million times
10^{9}	giga-	one billion times
10^{12}	tera-	one trillion

22

The International System of Units (système international d'unités), also known as SI Units, is used for measurements and calculations in chemistry. SI Units are based on the metric system and have been adopted by most countries. The following chart shows commonly used SI-based units.

SI UNITS

QUANTITY MEASURED	SYMBOL FOR UNIT	NAME OF UNIT
electric current	A	amperes
length	m	meter
mass	kg	kilogram
time	s	second
temperature	K	kelvin
volume	L	liters
heat	J	joule
amount of a substance	mol	mole

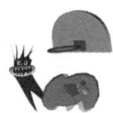

The Super-Charged World of Chemistry Part 1

0:15:39

SECTION B: UNCERTAINTY IN MEASUREMENT

When we measure a physical property and assign a numerical value to that measurement, the numerical value is always an approximation. Because the accuracy of our measurement depends on the accuracy of the instrument we use, and no instrument is absolutely reliable, the terms "precision" and "accuracy" can be as sticky as a wad of bubble gum on a hot day. You may think they mean the same thing, but nope! Not in science.

0:15:58

Precision measures the reproducibility of a result. In other words, if we take the same measurement five times and we get the same answer every time, our results are precise. **Accuracy** refers to how close our measurement is to the theoretical "true value." The "true value" is determined by some brilliant chemist and is listed in your text book.

(Here's an example. When you're frolicking in the woods with your crossbow, precision refers to your ability to hit wild, foliage-munching targets over and over, and accuracy refers to how close your shot is to a bull's eye.)

V1

`0:16:51` **SIGNIFICANT FIGURES**

When you decide how many digits you want to use to represent a value, you are deciding how many significant figures to use. Significant figures include the numbers on either side of the decimal point that seem reasonably certain.

For instance, the number 1.075 has four significant figures.

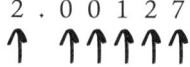

2.00127 has six significant figures.

$$2 \cdot 0\ 0\ 1\ 2\ 7$$
↑ ↑↑↑↑↑

VIDEO NOTES

The Super-Charged World of Chemistry Part 1

The number 0.0075 has just two significant figures, since the zeros in the number only serve to locate the decimal point.

$$0 \cdot 0 0 7 5$$

0.0000362 has only three significant figures. Those five zeros don't have value; they just tell you where the decimal point is.

$$0 \cdot 0 0 0 0 3 6 2$$

When you do calculations, you have to decide how many significant figures to use in your answer. If the answer has more figures than are significant, you can round off your answer.

Half double decaffeinated half-caf.
With a twist of lemon.

– **Steve Martin** in *L.A. Story*

V1

26

Here are some general rules for significant figures:

Addition & subtraction problems

Give the result to as many decimal places as the number in the problem with the fewest decimal places. Remember, decimal places are only the numbers after the decimal point.

10.712 + 3.61 = 14.322

= 14.32 rounded off to account for significant figures

Since the number 3.61 has only two decimal places, we use two decimal places in our answer and round to 14.32. The final 2 just isn't significant!

multiplication & division problems

Round off the result so it has the same number of significant figures as the number in the problem with the fewest significant figures.

Significant figures can be on either side of the decimal point.

2.32 + 77.96 = 180.8672

= 181 rounded off to account for significant figures

Since 2.32 only has three significant figures, our answer can only have three significant figures and we round up to 181. We handle division problems exactly the same way that we handle multiplication.

0:19:20 **SCIENTIFIC NOTATION: A SPLIT-SECOND REVIEW**

Scientific notation is a hip way to write really big or really small numbers. It cuts back on the decisions you have to make about significant figures, because it reduces the number of zeros in big numbers by expressing them in terms of powers of ten.

For example, instead of writing 32,100,000,000, you'd write 3.21×10^{10}. A small number like 0.000000084 would be written as 8.4×10^{-8}.

DEIONIZED INFORMATION:

For $0.000000084 = 8.4 \times 10^{-8}$, count the number of times you need to move the decimal point in 8.4 to get your original number, and then use that number as the exponent of 10. If you have to move the decimal point to the left, the exponent is negative; if you move it to the right, it's positive.

VIDEO NOTES

The Super-Charged World of Chemistry Part 1

SECTION C: DIMENSIONAL ANALYSIS

`0:20:20`

Although it sounds very complex, **dimensional analysis** is only a fancy term that describes the process of converting units. Chemistry uses lots of different units, like meters, liters, seconds, and grams. Sometimes you need a particular unit to do a calculation, and if the unit you need is not the one given in the problem, you'll have to use the unit factor method to convert units.

Agh! Dimensional Analysis!

`0:21:41`

To use the unit factor method, all you need to do is write out the units––kilograms, seconds, meters, or whatever––for every quantity in the calculation. Then carry the units through the calculation, treating them as algebraic quantities.

HERE'S AN EXAMPLE.

Willie the Wild Worm

wants to visit his girlfriend Lilly, who lives in a compost heap 7.92 miles away. Willie can only wiggle along one centimeter at a time. How many centimeters will poor, lovesick Willie have to wiggle to cover the 7.92 miles between him and his beloved? Willie measures his travel in centimeters, but we've measured the length of his trip in miles. We have to convert our miles into centimeters.

To use the unit factor method, multiply 7.92, the quantity you start with, by the appropriate conversion factor until you get the unit you want. In this case, since your answer should be in terms of centimeters, you have to convert miles to feet, then feet to inches, and finally inches to centimeters.

1 mile = 5,280 feet

1 foot = 12 inches

1 inch = 2.54 centimeters

The Super-Charged World of Chemistry Part 1

The conversion factor from miles to feet looks like this:

$$\frac{5{,}280 \text{ feet}}{1 \text{ mile}} = 1$$

5,280 feet is equal to 1 mile, so in our equation, multiplying by 5,280 feet over 1 mile is just like multiplying by 1. The conversion factors that convert feet to inches and inches to centimeters work the same way, so multiplying by each conversion factor is like multiplying by one.

$$1 = \frac{12 \text{ in.}}{\text{ft}}$$

$$1 = \frac{2.15 \text{ cm}}{4 \text{ in.}}$$

To find out how many centimeters Willie has to wiggle, set the problem up this way:

$$\text{number of cm} = 7.92 \text{ miles} \times \frac{5{,}280 \text{ ft}}{1 \text{ mile}} \times \frac{12 \text{ in.}}{1 \text{ ft}} \times \frac{2.45 \text{cm}}{1 \text{ in.}}$$

$$7.92 \text{ miles} \times \frac{5{,}280 \text{ ft}}{1 \text{ mile}} \times \frac{12 \text{ in.}}{1 \text{ ft}} \times \frac{2.45 \text{cm}}{1 \text{ in.}} = 1.27 \times 10^6 \text{cm}$$

Cross out the "miles," "feet," and "inches," and you get a product of 1.27×10^6 centimeters.

Since each of the conversion factors equals 1, we are not changing the value of the measurement at all. All we are changing is the units. When we cancel out the miles, feet, and inches in the problem and then multiply it through, we get 1.27×10^6 centimeters. Using these conversion factors is like multiplying 7.92 times 1 times 1 times 1, so that we manage to change only the units, not the values.

7.92 only has three significant figures, so the answer can only have three significant figures. With this in mind, we can say that Willy the Wild Worm must wiggle a hearty 1.27×10^6 centimeters to get to Lilly's compost heap.

Atomic Advice

· Using the unit factor method, you'll get the unit you want by carrying and cancelling units algebraically.

· Conversion factors are exact numbers and do not aVect the number of significant figures.

Quiz 1

(ANSWERS ON PAGE 296)

1. Classify the following as mixtures, compounds or elements.
   (CAUTION: This is not a recipe. If you are looking for a recipe, see the Stress Relief section for some finger-lickin' good stuff.)

    a. milk   _____

    b. baking soda   _____

    c. table salt   _____

    d. mercury   _____

    e. orange juice   _____

    f. ice cream   _____

    g. soda water   _____

    h. chlorine   _____

    i. air   _____

2. Complete the following mathematical expressions using appropriate number of significant figures.

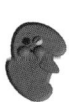

34

a. $7.91 - 2.51$ _____

b. $10.375 + 8.8$ _____

c. $2.73 \times 10^3 \times 9.1 \times 10^{-2}$ _____

d. $\dfrac{2.73 \times 10^3}{9.1 \times 10^{-2}}$ _____

e. $4.337 + 1.663$ _____

f. $5.887 \times 10^{-6} \times 2.111 \times 10^{10}$ _____

3. How many milliliters are in 2.07 gallons?

4. How many miles are there in 1.07×10^5 centimeters?

5. How many pounds are there in 8.72 nanograms?

The Super-Charged World of Chemistry Part 1

STOICHIOMETRY

`0:23:49`

Stoichiometry deals with the quantities of stuff in chemical reactions. We use stoichiometry to answer the question, "How much?"

STOICHIOMETRY

`0:24:13`

SECTION A: CHEMICAL EQUATIONS

A chemical equation is a short-hand way to describe a reaction. It gives the formulas for what's being mixed together (the reactants) and what you get (the products). The reactants are what's being mixed together, and the products are what you get.

V1

A SIMPLE CHEMICAL EQUATION:

reactants   products

$$A + B \rightarrow C + D$$ "$\rightarrow$" *means* **YIELDS**

`0:24:40`

A and B are the reactants, and C and D are the products. Chemical equations must be consistent with the **law of conservation of mass**, which says that atoms are neither created nor destroyed in chemical reactions. This means that the products of the reaction must *have the same total number of atoms of each element* as the reactants had before the reaction. The key thing to remember here is that the atoms in the reactants have only been rearranged, no atoms have been added or taken away.

`0:25:51` ## BALANCING EQUATIONS

To avoid breaking the law of conservation of mass, you have to learn to balance equations. In chemistry you will balance equations over and over again. It's just one of those things you have to learn. Before you do a calculation, always look at the equations you're given to make sure they are balanced. If your equations are not balanced, the rest of your calculation will be wrong.

The balanced chemical equation that represents the water-forming reaction of hydrogen and oxygen looks like this:

$$2H_2 + O_2 \rightarrow 2H_2O$$

H_2 is a molecule with 2 hydrogen atoms, and O_2 is a molecule with 2 oxygen atoms. Each water molecule has 2 hydrogen atoms and 1 oxygen atom. Here's the unbalanced equation that represents the formation of water:

$$H_2 + O_2 \rightarrow H_2O$$

The STOICHIOMETRY SISTERS SAY:

NO WATERLOGGING!

It's easier than balancing your checkbook in the bathtub.

This equation is unbalanced because there are 2 hydrogen atoms and 2 oxygen atoms on the left side, but 2 hydrogen atoms and only 1 oxygen atom on the right. The law of conservation of mass says that atoms are not created or destroyed in chemical reactions, so the equation above violates the law. It's up to us to rehabilitate that equation!

All right, let's balance that bad boy. Since there are 2 oxygen atoms on the right side of the equation, we have to make sure both atoms are accounted for in the product on the right side of the equation. To balance the number of oxygen atoms on both sides of the equation, we place a 2 in front of the product H_2O.

$$H_2 + O_2 \rightarrow\, = 2H_2O$$

Now there are 2 oxygen atoms on the left, and 2 oxygen atoms on the right, so the oxygen atoms are balanced. The hydrogen, however, still is not balanced. There are 2 hydrogen atoms on the left side, but now we have 4 hydrogen atoms on the right side. But all is not lost. Placing a 2 in front of the hydrogen molecule on the left side of the equation balances the hydrogens and saves the equation from THE ABYSS OF U$_N$B$_A$L$_A$$_NCE$$_D$ EQUATIONS.

$$2H_2 + O_2 \rightarrow 2H_2O$$

Ta-dah! That cleaned up nice and easy. Let's try another one.

VIDEO NOTES

The Super-Charged World of Chemistry Part 1

When iron (Fe) and oxygen (O) react with each other, iron oxide, or rust, is formed. The equation below says that iron and oxygen combine to yield a compound that has 2 iron atoms and 3 oxygen atoms.

$$Fe + O_2 \rightarrow Fe_2O_3$$

We need to balance the 2 oxygens on the left with the 3 oxygens on the right. Since 2 and 3 multiply together to get 6, we'll try to get 6 oxygen atoms on each side of the equation. A 3 in front of the O_2 on the left and a 2 in front of the Fe_2O_3 on the right will bring each side to a total of 6 oxygen atoms.

$$Fe + 3O_2 \rightarrow 2Fe_2O_3$$

Now that the oxygen is balanced, we'll go on to balance the iron. Since we put a 2 on the right side of the equation, there are now 4 iron atoms on the right. Getting 4 iron atoms on the left side of the equation is no biggie: just put a 4 in front of the Fe.

$$4Fe + 3O_2 \rightarrow 2Fe_2O_3$$

$$\rightarrow \text{rust}$$

V1

SECTION B: ATOMIC AND MOLECULAR MASS

`0:31:04`

Of all the elements, hydrogen (H) has the smallest mass and the simplest structure. A hydrogen atom consists of two electrically charged particles: a **proton**, which has a positive charge, and an **electron**, which has a negative charge. As small as it is, the hydrogen atom has measurable mass, and 99.9% of that mass is contained in the atom's **nucleus**, where the protons are.

All atoms except hydrogen have another type of particle in the nucleus called a **neutron**. Neutrons, which reside in the nucleus along with the protons, have no charge. They're as neutral as a doormat in a beige suit.

The nucleus of every atom has an electrical charge determined only by its protons, since neutrons have no charge. The neutrons definitely count, though, when you consider mass. The mass of the nucleus is made up of the combined masses of the protons and neutrons. The combined mass of the nucleus (protons and neutrons) and the electrons is the total **atomic mass.**

`0:32:49`

The mass number is the total number of protons and neutrons in the nucleus of an atom. Mass number is represented by the letter A. The **atomic number** equals the number of protons in the nucleus and is represented by the letter Z.

`0:33:18`

VIDEO NOTES

The Super-Charged World of Chemistry Part 1

Below is the general notation for representing the mass number and atomic number of any element.

$$_A^Z E$$

Here is what the notation looks like for the element helium. Helium has a mass number of 4, with 2 protons and 2 neutrons in its nucleus. Helium's atomic number, which represents only the number of protons, is 2.

$$_4^2 He$$

The periodic table has eighteen main vertical columns called *groups*, and seven main horizontal rows called *periods*. This arrangement places elements with similar properties in the same column. The average atomic mass is usually noted above the symbol of an element and the atomic number appears below the symbol.

I'm the first element with a neutron on the periodic table. My neutron increases my mass, gives me x-ray vision, and lets me fly at supersonic speeds!

V1

There are two general trends: going down any group, the atoms get larger, and going from left to right across any period, the atoms get smaller.

PERIODIC TABLE OF ELEMENTS

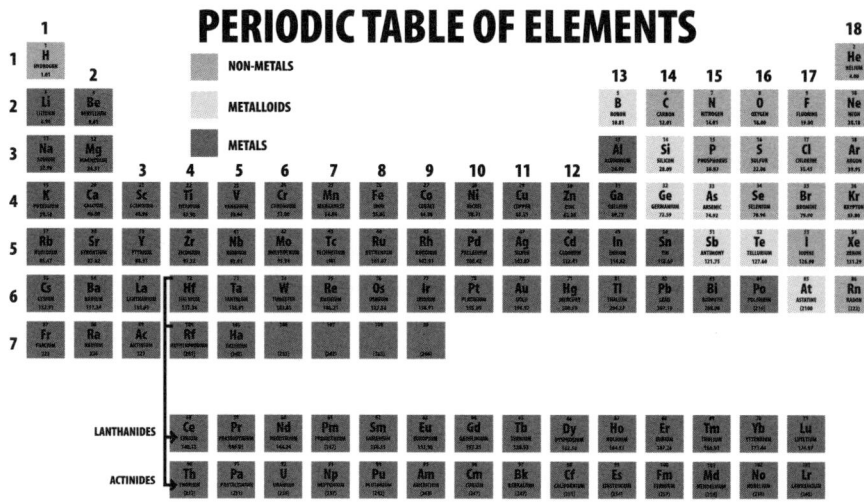

(For a full-size version of the periodic table, check out Other Important Stuff N⁰. 10.)

Some of an element's atoms have different numbers of neutrons in the nucleus than appear in the periodic table, since the periodic table just gives average masses. Such atoms are called **isotopes**.

The Super-Charged World of Chemistry Part 1

Take carbon, for example. On the left is carbon with a mass number of 12.

$$^{6}_{12}C \qquad ^{6}_{14}C$$

By looking at the atomic number and mass number of the carbon atom on the left, you can tell that it has 6 protons and 6 neutrons. You know this because the atomic number indicates that it has 6 protons, and the mass number indicates it has a total of 12 protons plus neutrons. You just subtract the number of protons from the mass number to figure out how many neutrons are in the atom's nucleus: 6 in this carbon isotope.

The number of neutrons in an atom's nucleus often varies within an element, so the mass of individual atoms can have more than one value.

On the right is carbon-14, another carbon isotope. It also has 6 protons, of course, but since the mass number is 14, we know that there must be 8 neutrons in the nucleus.

COMB YOUR MEMORY:

> **atomic number = smaller**
> **mass number = larger**

`0:35:51` Atomic masses are measured in atomic mass units (amu). All atomic mass units are based on the standard of carbon-12, and since carbon-12 is assigned a mass of 12 amu, that means one atomic mass unit has a mass equal to one-twelfth of a carbon-12 atom. Measuring the mass of atoms in terms of amu keeps the measurement darn close to the mass number, A.

I hate definitions.

– Benjamin Disraeli

CHEM SAFETY GUY says,

WARNING:

HANDLE WITH CARE.

HAZARDOUS TERMINOLOGY: Amu is sometimes referred to as *dalton*. Do not panic if you see the term dalton—just handle it the same way you would an amu.

NO WIGGING OUT.

46

`0:34:54` Some elements have several isotopes. We can calculate the average mass of an element's isotopes using their amu values. The mass in amu listed for each element in the periodic table is actually the average mass of the element's isotopes: it's the hypothetical "average" atom for each element.

To find the average mass of the element's isotopes, you need to know each isotope's atomic mass (no prob, they're listed in the periodic table), as well as each isotope's **natural abundance** (also listed in your textbook).

Natural abundance describes how much there is of a particular isotope in comparison with other isotopes of the same atom. For example, 99.985% of all the hydrogen in the world is the hydrogen isotope with one proton; so, the natural abundance of the hydrogen isotope with one proton is 99.985%.

Here's how to calculate the average mass of two of hydrogen's isotopes: Multiply the atomic mass of each isotope by its natural abundance, then add together both contributions.

isotope mass $\times$ natural abundance $=$ contribution

^{1}H 1.00783 amu 0.99985 $=$ 1.0077 amu

^{2}H 2.01410 amu 0.00015 $=$ 0.00030 amu

Total average atomic mass $=$ 1.0080 amu

The hydrogen isotope with 1 proton has a natural abundance of 99.985% (or 0.99985), and the isotope with 1 proton and 1 neutron, called deuterium, has a natural abundance of 0.015% (or 0.00015). The mass of the first isotope is 1.00783 amu and the mass of deuterium is 2.01410 amu. When we multiply the mass of each isotope by its natural abundance and add the two products together, we get the average atomic mass of hydrogen: 1.0080 amu.

GROOVE ON.

SECTION C: MOLES

`0:40:38`

Atoms and molecules are so extremely, unbelievably small that even the tiniest samples of matter contain unfathomably large numbers of them. To handle these tremendous numbers, chemists use a funky unit of measurement called the **mole**.

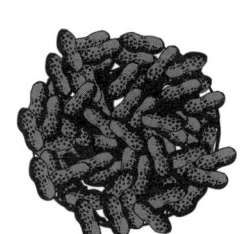

A mole is a unit of measurement that represents 6.022×10^{23} things, and should not be confused with the burrowing insectivore with bad eyesight. Moles are designed to measure atoms, molecules, **ions**, electrons, sesame seeds, and tiny things like that.

ONE MOLE OF BEER NUTS WOULD COVER THE ENTIRE SURFACE OF THE EARTH!

The Super-Charged World of Chemistry Part 1

Think of the term *mole* like the term *dozen*. You know that a dozen refers to twelve things: twelve eggs, twelve donuts, whatever. A *mole* refers to 6.022×10^{23} eggs or donuts or atoms or molecules or whatever you're measuring.

The magic mole number, 6.022×10^{23}, is also known as Avogadro's Constant (not to be confused with Avocado's Constant, which is a recipe for darn good guacamole—see page 69.)

Avogadro's magic number is 6.022×10^{23}...

$$6.022 \times 10^{23}$$

50

10^{23}

Like scientific notation, the mole makes it easier to deal with unwieldy numbers. One mole of atoms contains 6.022×10^{23} atoms; so one mole of water molecules is 6.022×10^{23} molecules. In that one mole of water molecules, there are 3 moles of atoms, since each water molecule consists of 3 atoms––2 hydrogens and 1 oxygen.

Here's another **cool thing** about **moles:** One mole of any type of atom has a mass in grams equal to that atom's *atomic mass measured in amu.*

When you look up the average atomic mass of carbon (C) in the periodic table and see that its atomic mass is 12.01 amu, you know instantly that one mole of carbon atoms weighs 12.01 grams. Iron (Fe) has an atomic mass of 55.847 amu, so one mole of iron atoms weighs 55.847 grams. Conversion from amu to grams per mole is that easy, which is a good thing, because chances are you'll be doing that conversion a lot!

amu

> **DEIONIZED INFORMATION:** Since one mole of any type of atom has a mass in grams equal to the atom's mass measured in amu, the numbers don't change when you convert your units to grams per mole.
>
> **For example:**
>
> 1 lithium atom weighs 6.941 amu
>
> $\rightarrow$ 1 mole of lithium atoms weighs 6.941 grams

`0:41:00`

`0:44:23`

Molecular mass represents the combined masses of the atoms that form a single molecule. Add up the masses (in amu) of the molecule's atoms, and you have the molecular mass. Molar mass equals the mass of one mole of a substance. When you do calculations, you will usually use molar mass in grams per mole (g/mol), which is how many grams one mole of your substance would weigh. For instance, the molar mass of carbon dioxide (CO_2) is 44.01 grams per mole, so if you had a mole of carbon dioxide molecules, it would weigh 44.01 grams.

Now let's traipse down the path to the molar mass of carbon dioxide (CO_2).

First off, we add the atomic mass of carbon (C) to 2 times the atomic mass of oxygen (O). Why 2 times the atomic mass of oxygen? Because there are 2 oxygen atoms in each carbon dioxide molecule.

Molar mass of C = 12.01 g/mol

Molar mass of O = 16.00 g/mol

$$1(12.01 \text{ g/mol}) + 2(16.00 \text{ g/mol}) = 12.01 \text{ g/mol} + 32.00 \text{ g/mol}$$
$$= 44.01 \text{ g/mol}$$

(the molar mass of CO_2)

Molar mass is *always* expressed in grams per mole (g/mol), since we are measuring an entire mole. The molar mass of water molecules equals the mass of 2 moles of hydrogen atoms plus the mass of 1 mole of oxygen atoms.

The Super-Charged World of Chemistry Part 1

We multiply the mass of hydrogen times 2 because there are 2 moles of hydrogen atoms in each mole of water molecules.

$$\text{Molar mass of H} = 1.0080 \text{ g/mol}$$

$$\text{Molar mass of O} = 16.00 \text{ g/mol}$$

$$(2 \times 1.0080 \text{ g/mol}) + 16.00 \text{ g/mol}$$
$$= 18.02 \text{ g/mol}$$

Water's molar mass, or the mass of one mole of water molecules, is 18.02 grams per mole.

Recipe (in its entirety) for boiled owl:

Take feathers off. Clean owl and put in cooking pot with lots of water. Add salt to taste.

– *The Eskimo Cookbook* (1952)

STUDY SIDEKICK

54

`0:49:34` **SECTION D: PERCENT COMPOSITION**

The percent composition of a compound gives the ratio of elements in the compound. The compound glucose ($C_6H_{12}O_6$) is made up of 6 carbon atoms (C), 12 hydrogen atoms (H), and 6 oxygen atoms (O). Some percentage of the glucose is carbon, some percentage is hydrogen, and the remainder is oxygen. Using another handy-dandy chem formula, we can figure out these percentages.

$$\text{percent composition} = \frac{\text{total mass of an element}}{\text{total mass of a compound}} \times 100\%$$

$$\text{percent composition} = \frac{(\text{\# of atoms of an element})(AW)}{\text{molar mass of the compound}} \times 100\%$$

UNC**L**E **E**TH**E**R'S
LAUNDRY PILE

30% jeans
20% t-shirts
5% handkerchiefs
10% underwear
10% denim overalls
25% knee-high socks

VIDEO NOTES

The Super-Charged World of Chemistry Part 1

First we determine the molar mass of glucose, expressing the mass in grams per mole. We just look up each element's weight in grams per mole in the periodic table, multiply by the number of atoms we have of that element, and add up all the values.

$$\text{molar mass of C} = 12.01 \text{ g/mol}$$
$$\text{molar mass of H} = 1.008 \text{ g/mol}$$
$$\text{molar mass of O} = 16.00 \text{ g/mol}$$

$$\text{Total molar mass} = (6 \times 12.01 \text{ g/mol}) + (12 \times 1.008 \text{ g/mol}) + (6 \times 16.00 \text{ g/mol}) = 180.16 \text{ g/mol}$$

The molar mass of glucose ($C_6H_{12}O_6$) is 180.16 grams per mole. Pretty slick, eh? The next time a stranger tries to give you a box of bonbons, remember to retort, "The molar mass of glucose is 180.16 grams per mole."

Now we're ready to calculate percent composition. First, we find the percentage of carbon using the equation above. Since there are 6 carbon atoms in one molecule of glucose, we take 6 times the atomic weight of carbon, which is 12.01 grams per mole, and divide by the molar mass of glucose, which we just figured out is 180.16 grams per mole. Next, we multiply by 100 to phrase our answer as a percentage.

$$\%C = \frac{6 \ (12.01 \text{ g/mol})}{180.16 \text{ g/mol} \times 100} = \frac{72.06 \text{ g/mol}}{180.16 \text{ g/mol} \times 100} = 40.00\%$$

V1

We do the same thing to calculate the percentage of hydrogen in glucose.

$$\%H = \frac{12(1.008 \text{g/mol})}{180.16 \text{ g/mol}} \times 100$$

$$= \frac{12.10 \text{ g/mol}}{180.16 \text{ g/mol}} \times 100 = 6.713\%$$

Just once more for the percentage of oxygen in glucose.

$$\%O = \frac{6(16.00 \text{g/mol})}{180.16 \text{ g/mol}} \times 100$$

$$= \frac{96.00 \text{ g/mol}}{180.16 \text{ g/mol}} \times 100$$

$$= 0.5327 \times 100 = 53.29\%$$

By this point, calculating percentages should be old hat --so here's something else for your calculator to crunch: You can double-check your work by adding all the percentages together to make sure they equal 100%.

Some of the many hats of Uncle Ether.

The Super-Charged World of Chemistry Part 1

SECTION E: EMPIRICAL FORMULAS

`0:54:21`

A compound's molecular formula tells you the exact number of each type of atom in one molecule of the compound. **Empirical formulas**, on the other hand, show ratios rather than exact quantities of atoms. For example, hydrogen peroxide's molecular formula, H_2O_2, indicates that each hydrogen peroxide molecule consists of 2 hydrogen atoms and 2 oxygen atoms. Hydrogen peroxide's empirical formula, HO, only tells you the ratio of hydrogen to oxygen in a hydrogen peroxide molecule.

   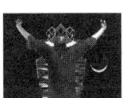  

"All the galaxy will kneel to the power of the Dark Side!"

The same holds true for glucose: its molecular formula, $C_6H_{12}O_6$, tells you that each glucose molecule consists of 6 carbon atoms, 12 oxygen atoms, and 6 oxygen atoms. The empirical formula for glucose, CH_2O, tells you only the ratios of atoms in a glucose molecule.

In many cases, a molecule's molecular formula is the same as its empirical formula, like in the case of carbon dioxide (CO_2). Each carbon dioxide molecule has 1 carbon atom and 2 oxygen atoms, and the ratio of oxygen to carbon is 2 to 1.

V1

> **ATOMIC ADVICE:** Atomic Weight (AW) is the mass of an atom. The average atomic mass for each element is indicated on the periodic table.
>
>
>
> | 1 |
> | H |
> | Hydrogen |
> | 1.008 |

Earlier on we discussed percent composition, the ratio of a compound's elements to the total compound. We can use information about the percent composition of a compound to find the compound's empirical formula. Let's look at an example.

We have 100 grams of water and we happen to know that water is 11.19% hydrogen and 88.81% oxygen. Since we have a 100-gram sample, conversion to grams is a breeze: 11.19% of 100 grams is 11.19 grams of hydrogen and 88.81% of 100 grams is 88.81 grams of oxygen. To determine the ratio of hydrogen to oxygen, we'll just compare how many moles we have of each element. We start off by figuring out the number of moles of hydrogen atoms we have, converting from grams to moles by the unit factor method.

$$\text{moles of hydrogen} = \frac{1 \text{ mol H}}{\text{AW of H}} \times \text{weight of sample of H}$$

$$= \frac{1 \text{ mol H}}{1.008 \text{ m g/mol}} \times 11.19 \text{ g H}$$

$$- 11.10 \text{ mol H}$$

The Super-Charged World of Chemistry Part 1

Now we'll convert the mass of oxygen to moles using the same method.

$$\text{moles of oxygen} = \frac{1 \text{ mol O}}{16.00 \text{ g/mol}} \times 88.81 \text{ g O} = 5.551 \text{ mol O}$$

The ratio of moles of hydrogen atoms to moles of oxygen atoms is $11.10 : 5.55$, or $2 : 1$. For every 2 hydrogen atoms, there is 1 oxygen atom in our compound, so water's empirical formula is H_2O.

Here's how we found the percent composition of water (H_2O).

• **First,** find the molar mass (MM) of water.

mass of H = 1.008 g/mol

mass of O = 16.00 g/mol

MM = 2(1.008g/mol) + 16.00 g/mol = 18.02 g/mol

• **Next,** use the percent composition equation.

$$\% \text{ composition} = \frac{(\text{number of atoms of an element})(\text{AW})}{\text{molar mass of the compound}} \times 100$$

$$\%H = \frac{2(1.008 \text{ g/mol})}{18.02\text{g/mol}} \times 100 = 11.19\%$$

$$\%O = \frac{16.00\text{g/mol}}{18.02\text{g/mol}} \times 100 = 88.79\%$$

Empirical formulas are also useful when we work with substances that do not consist of molecules. Obviously, we cannot calculate a molecular mass for substances that have no molecules. That would be like counting the number of purple jelly beans in a bag of gummy worms. Instead, we calculate a **formula weight**, which is simply the sum of the masses of the atoms in the empirical formula.

Sodium chloride (NaCl) is a crystal, not a molecule, so we have to calculate its formula weight. Crystals are different from molecules, because crystals do not have definite pairing and bonding of atoms or ions like molecules do. Crystals are more like a bunch of atoms stacked up on one another like the bricks in a building. The atoms or ions in a crystal stack up in a repeating pattern, forming a **crystal lattice**.

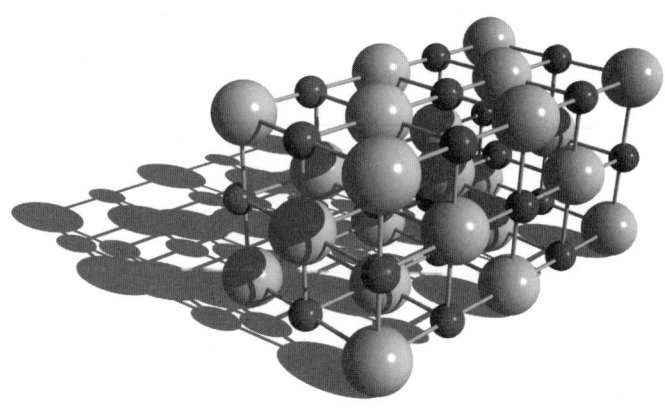

The *STOICHIOMETRY SISTERS SAY:*

It's easier than chuting down Niagra Falls in a handbag.

Formula weight and molar mass are almost the same thing, but formula weight is the weight of atomic structures that are not molecules. Finding formula weight is very easy. The process is the same as you used to find molar mass, but you base your calculation on the number of atoms indicated by the structure's empirical formula, rather than on a molecular formula.

To find the formula weight of sodium chloride (NaCl), add up the masses of sodium (Na) and chlorine (Cl).

$$\text{molar mass of Na} = 22.9898 \text{ g/mol}$$

$$\text{molar mass of Cl} = 35.453 \text{ g/mol}$$

$$\text{Formula weight} = 22.9898 \text{ g/mol} + 35.453 \text{g/mol}$$

$$= 58.4428 \text{ g/mol}$$

$$= 58.443 \text{ g/mol (significant figures)}$$

Remember the rules for significant figures with addition and subtraction. Your answer can only have as many decimal places as the term in the problem with the fewest decimal places. The mass of chlorine, one of the terms we added, only has 3 decimal places, so our answer can only have 3 decimal places.

62

SECTION F: PROBLEMS BASED ON CHEMICAL EQUATIONS

To calculate the masses of substances taking part in a chemical reaction, follow these four steps.

1. Write the balanced equation for the reaction.

2. Convert the known mass of one of the reactants or products into moles.

3. Use the balanced equation to set up the appropriate conversion factors, so you can find the number of moles of the other reactants or products.

4. Convert your values from moles back to grams.

COOKING WITH…
PROFESSOR
ROWLEY!

Professor Rowley, our incomparable culinary chemist, whipped up a tempting batch of magnesium oxide (MgO). Let's do a problem based on the recipe (a.k.a. chemical equation). If Professor Rowley starts with 2.4 grams of magnesium

(Mg) and an excess amount of oxygen (O) floating around in genuine laboratory air, how many grams of magnesium oxide (MgO) will he produce from this reaction? To figure it out, we'll follow the four steps listed above. The overall equation for the reaction is:

$$Mg + O_2 \rightarrow MgO$$

STEP 1: Write the balanced equation for the reaction.

$$2Mg(s) + O_2(g) \rightarrow 2MgO(s)$$

The equation tells us that 2 moles of magnesium reacting with 1 mole of oxygen yields 2 moles of magnesium oxide. Therefore, the number of moles of magnesium we start with will equal the number of moles of magnesium oxide we end up with.

> Rock and roll is the hamburger that ate the world.
>
> – Peter York

STEP 2: Convert the known mass of one of the reactants to moles.

$$AW = \text{atomic weight in grams per mole (g/mol)}$$

$$\text{moles of Mg} = \frac{1 \text{ mol Mg}}{AW \text{ of Mg}} \times \text{weight of sample (in grams)}$$

$$= \frac{1 \text{ mol Mg}}{24.0 \text{ g}} \times 2.4 \text{ g}$$

$$= 0.10 \text{ mol Mg}$$

STEP 3: Find how many moles of the product (magnesium oxide) there are by setting up a conversion factor. The conversion factor, which is based on information from the balanced equation, will convert moles of magnesium into moles of magnesium oxide without changing the number of moles. (We don't want to change the number of moles because the balanced equation shows that the number of moles of charbroiled magnesium oxide should equal the number of moles of magnesium.)

$$\text{mol MgO} = \text{mol Mg} \times \frac{2 \text{ mol MgO}}{2 \text{ mol Mg}}$$

Multiply the number of moles of magnesium, 0.10 moles, by 2 moles of magnesium oxide over 2 moles of magnesium (this is virtually the same as multiplying by one).

$$= 0.10 \text{ mol Mg} \times \frac{2 \text{ mol MgO}}{2 \text{ mol Mg}}$$

$$= 0.10 \text{ mol MgO}$$

STEP 4: Convert from moles back to grams. Multiply the number of moles of magnesium oxide, 0.10, times the formula mass (FM) of magnesium oxide.

$$\text{mass in grams of MgO} = \text{mol MgO} \times (\text{FM})\text{MgO}$$

$$= 0.10 \text{ mol} \times 40.3 \text{ g/mol}$$

$$= 4.0 \text{ grams MgO}$$

And voilà! Professor Rowley produces 4.0 grams of yummy magnesium oxide from 2.4 grams of magnesium.

Quiz 2

(ANSWERS ON PAGE 298)

1. Balance the following equations.

a. $C_5H_{12} + 8O_2 \rightarrow 5CO_2 + 6H_2O$

$C_5H_{12} + 8O_2 \rightarrow 5CO_2 + 6H_2O$

b. $AlCl_3 + 3H_2O \rightarrow Al(OH)_3 + 3HCl$

$AlCl_3 + 3H_2O \rightarrow Al(OH)_3 + 3HCl$

c. $Cl_2O_7 + H_2O \rightarrow HClO_7$

d. $NO_2 + H_2O \rightarrow HNO_3 + NO$

$3NO_2 + H_2O \rightarrow 2HNO_3 + NO$

e. $Fe_3O_4 + 4H_2 \rightarrow Fe + 4H_2O$

$Fe_3O_4 + 4H_2 \rightarrow 3Fe + 4H_2O$

2. Calculate the average atomic weight of chlorine from the following data.

Isotope	Isotopic mass	Relative abundance
^{35}Cl	34.96885	75.771%
^{37}Cl	36.96590	24.229%

$= 26.496$

$8.956 46 79$

35.453 g/mol 35.452719

3. What is the molar mass of the following compounds?

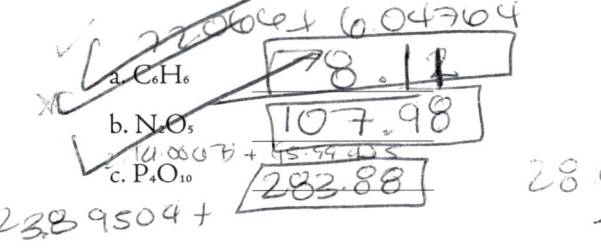

a. C_6H_6

72.0606 + 6.04764 *78.11*

b. N_2O_5

107.98

14.0067 + 15.9994·5

c. P_4O_{10}

283.88

123.89504 +

28.0134 +

79.97

4. If 7.82 grams represent 0.139 moles of an unknown compound, what is the formula mass of the compound?

M/n   *7.82 / .139*   *563*

56.26

5. Calculate the percent composition of each element in the following compounds.

a. SO_3 →*32.066* →*47.9982* *S = 40%* *O = 80.06* *80.06*

b. $Co(NO_3)_2$ *CO = 32%* *N = 15.3%* *60 00%* *6 = 52.9*

Co = 58.9332 +

N 28.01

14.0067 + 319

for 3 82.9396

N=30% 69.79

=32%

N=15%

52%

0.9599

6. Adipic acid has a percent composition of 49.3% C, 6.9% H and 43.8% O. What is its empirical formula?

14.0067

assume 100 g sample

C

(4) → $C_6H_{10}O_4$ *5.9*

1.5 $H_{2.5}O$ 49.3

O

6.9   *43.8*

12.011   *100799*   *15.9554*

4.1   *6.8*   *2.7*

2.7   *2.7*   *2.7*

1.5 (1.5   (3) < 2.5)

V1

68

7. Iron (Fe) reacts with oxygen (O_2) to produce iron oxide (Fe_2O_3). How many moles of iron oxide are produced from 0.150 moles of oxygen? How many grams of iron oxide are produced? (A Reminder from the Stoichiometry Sisters: Use the four steps for doing a calculation based on chemical equations.)

.150 mols of O.   ? mol

8. Dinitrogen pentoxide (N_2O_5) decomposes to nitrogen dioxide (NO_2) and oxygen (O_2). If a 2.621 gram sample of N_2O_5 is decomposed, how much NO_2 is formed?

$2 N_2O_5 \rightarrow 4 NO_2 + O_2$

2.621 =   2.621 × mm

? NO₂ =

2.621 / mm
.0502 mol   4

2 mol
N₂O₅

.064
mol
of
NO2

.064 × mm

The **STOICHIOMETRY SISTERS SAY:**

It's easier on your nerves than the scent of a gingko tree on a sultry summer's eve.

2.949 g of NO₂

A V O C A D O ' S   C O N S T A N T

(YUMMY GUACAMOLE)

1 large, ripe avocado

1 tomato, chopped

1 small onion, chopped

2 tablespoons lemon juice

1 tablespoon vegetable oil

onion powder

garlic powder

salt

pepper

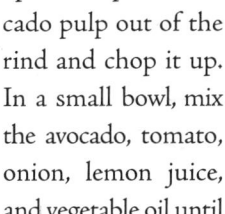

Cut the avocado in half and remove the pit. Scoop the avocado pulp out of the rind and chop it up. In a small bowl, mix the avocado, tomato, onion, lemon juice, and vegetable oil until creamy. Season with onion powder, garlic powder, salt, and pepper to taste.

A Tip for Your **Tropical Treat**: Put the avocado pit in the container with the guacamole. That way your concoction will stay delicious-looking and won't turn brown.

Variation: For peanut butter guacamole, add ½ cup peanut butter to the recipe above.

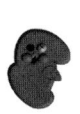

70

Solution Stoichiometry

A solution is a homogeneous mixture of two or more substances. In all solutions, there is a solute and a solvent. The largest quantity of substance present is usually referred to as the solvent, and the other substances are referred to as solutes. In a blue dye solution, the water is the solvent and the dye is the solute. In a solution that consists of a solid in a liquid, the solid is always called the solute and the liquid is always called the solvent.

A SOLUTE IS…

VIDEO NOTES

The Super-Charged World of Chemistry Part 1

SECTION A: MOLARITY

`1:10:34`

"**M**" stands for molarity. The molarity of a solution is the number of moles of solute in one liter of solution. If we say we have a 2-molar solution of sodium chloride in water, we mean that there are 2 moles of sodium chloride per liter of solution. Here is the equation used to find molarity (M).

$$M = \frac{n(\text{moles solute})}{V(\text{volume of solution in liters})}$$

M = molarity

n = number of moles

V = volume of solution in liters

The LETTER M STANDS for...

mallard   marshmallow   milk

V1

You can also use the volume and molarity of a solution to find the number of moles of solute you have. All you need to do is rearrange the molarity equation.

$$n = M \times V$$

LET'S LOOK AT A PROBLEM:

How many milliliters of a 0.400 M solution of sodium hydroxide do we need to react with 0.500 liters of 0.200 molar hydrochloric acid?

To work out the problem, we will use the molarity equations and a new variation to find volume. We modify the molarity equation to show that volume equals the number of moles divided by the molarity.

$$V = \frac{n}{M}$$

Mole of Knowledge:

"AQ" STANDS FOR "AQUEOUS," WHICH MEANS WATER IS THE SOLVENT

Here is the balanced chemical equation:

$$NaOH(aq) + HCl(aq) \rightarrow H_2O + NaCl(aq)$$

The balanced equation not only shows which chemicals are reacting to form what product, it also shows the ratio of reactants you need for the reaction to proceed so you have no reactants left over. According to the balanced chemical equation, sodium hydroxide (NaOH) and hydrochloric acid (HCl) react in a one-to-one mole ratio to form water (H_2O) and sodium chloride (NaCl). This means that for each mole of HCl we use, we need one mole of NaOH.

We start by determining how many moles of hydrochloric acid (HCl) there are in our 0.500 liter sample of hydrochloric acid. When we know this number, we will know the number of moles of NaOH we need. We use the rearranged molarity equation: $n = M \times V$.

The molarity of the hydrochloric acid solution is 0.200 moles per liter (mol/L). When we multiply 0.200 moles per liter by the volume of HCL (0.500 liters) we get 0.100 moles. Now we know there are 0.100 moles of HCl in the 0.500 liter sample.

$$n_{HCL} = M_{HCL} \times V_{HCL}$$

$$M = \frac{0.200 \text{ mol HCl}}{L}$$

$$V = 0.500 \text{ L HCl}$$

$$n_{HCL} = \frac{0.200 \text{ mol HCl}}{L} \times 0.500 \text{ L HCl}$$

$$= 0.100 \text{ mol HCl}$$

We can determine the number of moles of NaOH we need by looking at the balanced equation.

$$NaOH(aq) + HCl(aq) \rightarrow H2O + NaCl(aq)$$

This equation shows that there should be the same number of moles of NaOH as HCl, so we will need 0.100 moles of NaOH. We are now ready to calculate the necessary volume of NaOH, using the equation below.

$$V = \frac{n}{M}$$

$$V = \frac{0.001 \text{ mol NaOH}}{0.400 \text{ mol NaOH/L}}$$

$$V = 0.250 \text{ L of NaOH}$$

And that's all there is to it! We have our answer: we need 0.250 liters, or 250 milliliters, of sodium hydroxide to react with 0.500 liters of 0.200 molar hydrochloric acid. **Woohoo.**

`1:14:35` **SECTION B: DILUTIONS**

Dilutions change the concentration of solutions. When you add more solvent to a solution, it produces a solution of lower concentration. Take our old friend the blue dye solution, for example. Adding more water dilutes the little bugger, making the water a lighter blue and creating a solution with a higher percentage of water and a lower percentage of dye. We can find the new, post-dilution concentration of the solution with this simple equation.

$$M_1V_1 = M_2V_2$$

You might see the same equation written with C's and D's as the subscripts.

$$M_cV_c = M_dV_d$$

Don't freak, this equation means the same darn thing as the equation with numbers as subscripts--the c's stand for "concentrated solution" (your original solution) and the d's stand for "diluted solution." With this knowledge, you have foiled the demons of Chemistry Chaos. They will not succeed in bewildering you with their freakish letters.

The Super-Charged World of Chemistry Part 1

How would you prepare 250 milliliters of a 0.25 molar potassium nitrate solution from a stock solution of potassium nitrate that has a concentration of 1 mole per liter (1.00M)?

Here's a problem involving volume and concentration. The star of the show is the illustrious potassium nitrate (KNO_3).

In other words, we have a 1.00M KNO_3 solution and we want 250 mLs of 0.25M KNO_3 solution. How do we prepare it? That's right, sports fans, we use the equation we were just discussing.

$$M_1V_1 = M_2V_2$$

We can rearrange the equation to determine the volume of the stock solution we're trying to get.

$$V_1 = \frac{M_2V_2}{M_1}$$

$$V_1 = \frac{0.250M\ KNO_3 \times 0.250\ L\ KNO_3}{1.00M\ KNO_3}$$

$$= 0.0625\ L\ or\ 62.5\ mL$$

No problem! Our answer tells us that we need 62.5 milliliters of the stock solution to create 250 milliliters of 0.250-molar potassium nitrate solution. We prepare this tasty solution by taking 62.5 milliliters of the stock solution and adding just enough water to make 250 milliliters of solution.

V1

SECTION C: TITRATIONS

You probably know what **acids** are. Some sour-tasting foods, like orange juice and vinegar, are weak acids. Acids contain hydrogen ions (H^+); the more hydrogen ions in the acid, the stronger the acid. Bases contain hydroxide ions (OH^-); the more hydroxide ions, the stronger the base.

When an acid's hydrogen ions meet a base's hydroxide ions, the acid and the base neutralize each other. Similarly, when the right proportion of a base is mixed with an acid so that the amount of acid equals the amount of base, the mixture becomes neutral. The reaction's balanced chemical equation will tell you the ratio of starting chemicals you need for the reaction to yield a neutral solution.

$$NaOH + HCl \rightarrow H_2O + NaCl$$

This balanced equation indicates that one mole of sodium hydroxide (NaOH) reacts with one mole of hydrochloric acid (HCl) to produce one mole of water (H_2O) and one mole of sodium chloride (NaCl).

A **titration** is another way to determine the concentration of a solution. To find a concentration by titration, add a solution of the *unknown* concentration to a solution with a *known* concentration until the mixture reaches the equivalence point. The **equivalence point** is the point at which you have just enough base to neutralize all the acid in the solution, according to the balanced chemical equation. Dyes known as **acid-base indicators** can help you determine when the equivalence point has been reached.

Phenolphthalein is an acid-base indicator that turns from colorless to pink when the solution becomes neutral. The point at which the indicator changes color is called the **endpoint.** Not all acid-base indicators work the same way, so be sure to choose an indicator whose endpoint matches the equivalence point.

What a time! What a civilization!

– Cicero (106-43 BC)

V1

After giving one of his popular lectures on the delicate art of preparing flaky pie crusts, Professor Rowley demonstrated the proper technique for performing a titration. He started with a flask containing phenolphthalein and added 25.0 milliliters of a 0.15 M hydrochloric acid solution. He then added a sodium hydroxide solution of an unknown concentration until the hydrochloric acid solution turned pink. The pink tone signaled that the solution had been neutralized. At this point, Professor Rowley noticed that it had taken 35.0 milliliters of sodium hydroxide solution to neutralize the hydrochloric acid.

But--**horrors!**--our incomparable culinary chemist was still missing one essential piece of information: he didn't know the concentration of the NaOH solution.

Let's restate the problem as you might see it in class.

What is the concentration of an unknown NaOH solution if, during a titration, you need 35.0 mL of the NaOH solution to reach the equivalence point of 25.0 mL of a 0.150 M hydrochloric acid solution?

We'll start with the balanced equation that represents the reaction, since it's a useful way to tell the proportions of reactants and products in the reaction.

$$NaOH + HCl \rightarrow H_2O + NaCl$$

According to the equation, NaOH and HCl react in a one-to-one ratio, so the number of moles of NaOH will equal the number of moles of HCl. We started with 250 milliliters (0.025 liters) of 0.150-molar HCl. What concentration (molarity) would our NaOH solution have to be for it to react in a one-to-one ratio with 0.025 liters of hydrochloric acid? Here's the formula for molarity.

$$M = \frac{n}{V}$$

$$M = \text{molarity}$$
$$n = \text{number of moles}$$
$$V = \text{volume in liters}$$

$$M_{NaOH} = \frac{n_{NaOH}}{V_{NaOH}}$$

We can fill in for the variable V, because we know the volume of NaOH is 35.0 mL or 0.0350 liters; but we need to find n, the number of moles. Now what? Well, first we will find the number of moles of HCl neutralized by the NaOH. Then, since we know from the balanced equation that the number of moles of NaOH will equal the number of moles of HCl, we can calculate the number of moles of NaOH.

$$n_{NaOH} = n_{HCL}$$

$$n_{HCL} = M \times V$$

$$M_{HCL} = 0.150 \text{ M}$$

$$V_{HCL} = 0.025 \text{ L}$$

$$n_{HCL} = 0.150M \times 0.025 \text{ L}$$

$$n_{HCL} = 0.00375 \text{ mol}$$

Because the coefficients are equal in the balanced equation, the number of moles of NaOH will equal the number of moles of HCl--so there are 0.00375 moles of NaOH. Now we'll help Dr. Rowley finish off this beast by using the molarity formula to find the molarity of the NaOH solution.

$$M = \frac{n}{V}$$

$$M_{NaOH} = \frac{0.00375 \text{ mol}}{.035L}$$

$$M_{NaOH} = 0.107M \text{ or } 0.11M$$

REMEMBER significant figures!

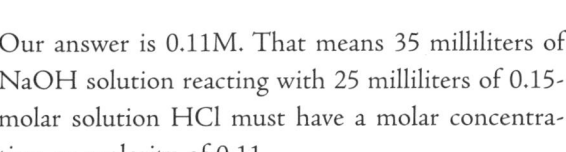

Our answer is 0.11M. That means 35 milliliters of NaOH solution reacting with 25 milliliters of 0.15-molar solution HCl must have a molar concentration, or molarity, of 0.11.

Now that Dr. Rowley has found the answer he sought, it is time for a limiting reagents fiesta!

84

`1:29:50` **SECTION D: LIMITING REAGENTS**

Reagents are the substances or compounds that take part in a chemical reaction. The limiting reagent is the one you run out of first––that is, the one that **HALT!** makes the reaction stop working.

Reagents should not be confused with the Hyatt Regency, the Hyatt hotel in downtown Atlanta. You know, the one that has a rotating, blue rooftop restaurant and is located next to the Peachtree Plaza Hotel, which is the tallest hotel in the world and the one from which the villain is pushed in the 1970's Burt Reynolds film, *Sharky's Machine*.

Let's take a closer look at how limiting reagents might elbow their way into one of your professor's questions:

There are an estimated 29,382,509 people with the last name *Smith* in the United States.

– *The Guiness Book of Records*

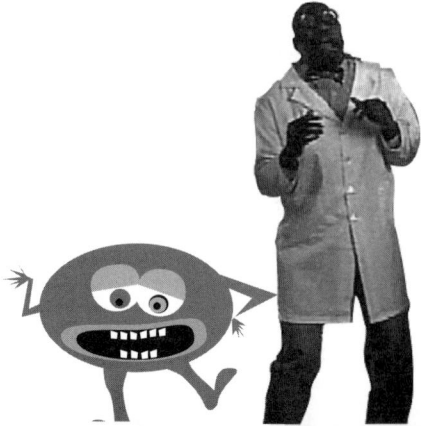

When 12.2 grams of magnesium (Mg) react with 24.0 grams of sulfur (S) to produce magnesium sulfide (MgS), what is the limiting reagent in the reaction?

$$Mg(s) + S \rightarrow MgS(s)$$

The way to tackle this one is to convert the amount of each reactant from grams to moles. Using the unit factor method to convert the 12.2 grams of magnesium and 24.0 grams of sulfur to moles, we get the following results:

$$\text{molar mass of Mg} = 24.3 \text{ g/mol}$$

$$\text{molar mass of S} = 32.0 \text{ g/mol}$$

$$AW = \text{atomic weight in grams per mole (g/mol)}$$

$$\text{moles of Mg} = \frac{1 \text{ mol Mg}}{AW \text{ of Mg}} \times \text{mass of Mg (in grams)}$$

$$\text{moles of Mg} = \frac{1 \text{ mol Mg}}{24.3 \text{ g}} \times 12.2 \text{ g} = 0.502 \text{ mol}$$

$$\text{moles of S} = \frac{1 \text{ mol S}}{AW \text{ of S}} \times \text{mass of S (g)}$$

$$\text{moles of S} = \frac{1 \text{ mol S}}{32.0 \text{ g}} \times 24.0 \text{ g} = 0.750 \text{ moles}$$

The balanced chemical equation indicates that the reactants will react in a one-to-one mole ratio.

$$Mg \text{ (s)} + S \rightarrow MgS \text{ (s)}$$

Since there are 0.750 moles of sulfur and only 0.502 moles of magnesium, there is more sulfur than can react with all the magnesium present. The amount of magnesium limits the reaction, so the limiting reagent is…yup, you guessed it: magnesium.

Now, we find the mass of the excess sulfur. The moles of excess sulfur will equal the amount you start with (0.75 moles) minus the amount that reacted with the Mg (0.502 moles, adjusted for significant figures to 0.50 moles).

$$0.75 \text{ mol} - 0.50 \text{ mol} = 0.25 \text{ mol of excess S}$$

The Super-Charged World of Chemistry Part 1

Okey dokey. We convert from moles back to grams by multiplying the moles of excess sulfur times the atomic weight of sulfur, all over 1 mole of sulfur.

mass in grams of sulfur = moles of excess sulfur $\times$ AW of S(s)

$$= 0.25 \text{ mol S} \times \frac{32.0 \text{g S}}{1 \text{mol S}}$$

$$= 8 \text{ grams S}$$

That's our answer: 8 grams of excess sulfur. We sure got that puppy licked.

`1:38:02` ## SECTION E: YIELDS

The term **yield** refers to the amount of product you get from a reaction. There are three different kinds of yields: theoretical yield, actual yield, and percent yield. The **theoretical yield** is the amount of product you predict you'll get from a reaction, based on its balanced equation and the limiting reagent. The **actual yield** is the amount of the product you actually get from an experiment. The **percent yield** is the actual yield (what you actually got) divided by the theoretical yield (what you were supposed to get), times one hundred.

$$\text{percent yield} = \frac{\text{actual yield}}{\text{theoretical yield}} \times 100$$

Let's look at a concrete example. When 1.00 mole of methane (CH_4) burns, it produces 1.75 moles of water (H_2O). What is the percent yield of water in the reaction? Using the balanced chemical equation and the amount of starting material, we can calculate the theoretical yield.

$$CH_4\,(g) + 2O_2\,(g) \rightarrow CO_2\,(g) + 2H_2O\,(l)$$

The Super-Charged World of Chemistry Part 1

The equation tells us that, theoretically, we should get 2 moles of water from every mole of methane burned: 2.00 moles is our theoretical yield. Using the equation for percent yield, we do a bit of math and obtain 87.5% for our percent yield.

actual yield = 1.75 moles H_2O

theoretical yield = 2.00 moles H_2O

$$\text{percent yield} = \frac{1.75 \text{ moles } H_2O}{2.00 \text{ mol } H_2O} \times 100 = 0.875 \times 100 = 87.5\%$$

A closed mouth gathers no feet.

– Unknown

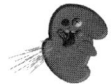

STUDY SIDEKICK

QUIZ 3

(A N S W E R S   O N   P A G E   3 0 1)

1. What is the molar concentration of each of the following solutions?

    a. 2.96 grams of potassium nitrate (KNO_3) dissolved in a total volume of 250.0 mL.

    b. 0.350 moles of calcium hydroxide, $Ca(OH)_2$, diluted to 1.67 liters.

2. You are given a solution of 12.0 M HCl. How much of the solution will you need to prepare 250.0 mL of a 1.50 M HCl solution?

3. If you titrate sulfuric acid (H_2SO_4) with sodium hydroxide (NaOH), how much 0.100 M NaOH is required to react with 31.5 mLs of 0.0750 M H_2SO_4?

VIDEO NOTES

The Super-Charged World of Chemistry Part 1

4. Aluminum (Al) metal reacts with oxygen (O_2) to produce aluminum oxide (Al_2O_3). How many moles of Al_2O_3 are produced when 0.100 moles of Al react with 0.100 moles of O_2?

_____

5. The equation below is the unbalanced equation for the reaction of chlorine and phosphorus:

$$Cl_2 + P_4 \rightarrow PCl_3$$

If 21.0 grams of Cl_2 are mixed with 53.2 grams of phosphorus, how much PCl_3 is produced? What is the limiting reagent? How much mass of the non-limiting reagent remains?

_____

6. An 18.4 gram sample of benzene (C_6H_6) reacts with excess nitric acid (HNO_3) to produce 19.5 grams of nitrobenzene ($C_6H_5NO_2$). What is the percent yield of nitrobenzene in this reaction?

$$C_6H_6 + HNO_3 \rightarrow C_6H_5NO_2$$

_____

Helium Man's
Super-Charged
Linguine with
Clam Sauce

1 lb. linguine

4 T. olive oil

2 T. chopped garlic (or less, to taste––but Helium Man thinks the garlic is the best part)

1½ c. minced canned clams

1 c. clam juice

⅓ c. dried parsley flakes

pepper

grated parmesan cheese

Cook the linguine according to package directions. In a medium-sized sauce pan, sauté the garlic in the olive oil until the garlic is lightly browned. (Careful not to burn it!) Add the clam juice, clams, parsley, and pepper to taste. Cover and simmer over low heat for 10-15 minutes. Pour the clam sauce over the cooked, drained linguine, sprinkle with grated parmesan cheese, and serve immediately.

Helium Man serves garlic bread and Caesar salad with linguine. (What a chef. What a superhero.)

CHEM SAFETY GUY says,

HEY YOU!

Flip to Practice Exam 1 in the Test Yourself section to tickle your brain and put your knowledge to work!

OUR EXAMS ARE CRUNCHY AND NUTRITIOUS.

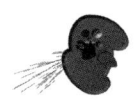

VIDEO TIME CODE

The Super-Charged World of Chemistry Part 2

Oh, this age! How tasteless and ill-bred it is!

– Catullus (ca. 84–54 BC)

96   THERMOCHEMISTRY

`0:03:12`

Thermochemistry measures the heat changes associated with chemical reactions. In other words, when energy moves around as a result of a reaction, thermochemistry figures out how much energy is involved and where the energy will end up. We measure energy in joules, which is an SI unit of measurement.

(The study of thermos chemistry deals with heat changes associated with the quantity and variety of soup stored in your handy beverage/soup container, and measures the variables chunkiness, saltiness, and The Steam Factor.)

To get cranking on our discussion of thermochemistry, we need to set some ground rules. So stick your nose nice and close to the page for a second: when you see **system**, think of all the substances taking part in a reaction, plus the reaction vessel. When you see **surroundings**, you'll know we're referring to the rest of the universe. Excellent, secret agents. We are now prepared to embark on our enthalpy expedition.

The Super-Charged World of Chemistry Part 2

SECTION A: ENTHALPY

Chemists use Greek letters to stand for words in mathematical equations. The Greek letter delta, Δ, which looks like your run-of-the-mill triangle, represents *change*. When you see delta in an equation, read it as "the change in" something; so when you see ΔH, read it as "the change in heat." The way you calculate the change in a quantity is by subtracting the initial value from the final value.

> *Citizens and statespeople alike simply subtract the initial value from the final value to get Δ.*

V2

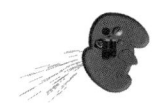

Enthalpy refers to the change in heat in a system at constant pressure. That is, enthalpy deals with the heat that flows in and out of a system as a result of a reaction.

$$\Delta H_{\text{PRODUCTS}} = \text{sum of enthalpies of the products}$$
$$\Delta H_{\text{REACTANTS}} = \text{sum of enthalpies of the reactants}$$
$$\Delta H_{\text{REACTANTS}} = \Delta H_{\text{PRODUCTS}} - \Delta H_{\text{REACTANTS}}$$

The equation above represents the enthalpy change during a reaction. ΔH is equal to the enthalpy of the products minus the enthalpy of the reactants.

There are two types of reactions that involve the exchange of heat: endothermic reactions and exothermic reactions. In an **endothermic reaction**, heat flows *into* the system. The change in an endothermic reaction's heat (ΔH) will always be a positive number, because heat is added as it flows from the outside surroundings into the reaction system.

`0:15:04`

When the opposite occurs and heat flows *out* of a system, you've got an **exothermic reaction**. In this case, you have a negative heat flow, because heat is exiting the system; so when you calculate the ΔH for an exothermic reaction, you'll always end up with a negative number.

`0:15:30`

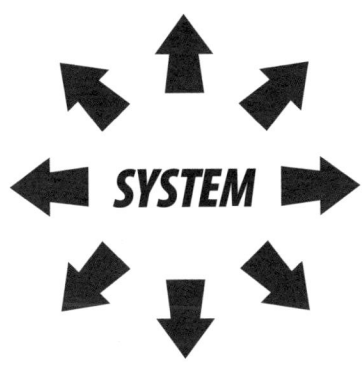

$\Delta H_f°$, **the standard enthalpy of formation,** is the enthalpy change in a reaction that forms a compound out of elements in their standard states. We can say that a substance is in its **standard state** only under the following conditions: the substance must be in its most stable form, at 25°C, and at one atmosphere of pressure. *Pressure* refers to the pressure that air exerts on a substance, and an *atmosphere*, which is abbreviated as *atm*, is a unit used to measure pressure.

`0:17:18`

V2

Atmospheric Factors that Increase Pressure:

1. HIGH GRAVITY

2. ZITS

3. SLEEP DEPRIVATION

4. DORM INVADED BY CHEESE-PIZZA-EATING MOOSE

5. ROOMMATE SNORES LIKE A MACK TRUCK REVVING UP

6. YOUR PRINT-OUT LOOKS LIKE THIS: ÇÅŒµ¥######

7. YOUR PROFESSOR LIKES TO PUT LATE PAPERS IN A MEAT GRINDER

8. YOUR BELLY BUTTON COMES UNDONE

9. THREE EXAMS ON THE SAME DAY

10. A COUP D'ETAT

11. AN UNFRIENDLY SPECTRE OF JOHN QUINCY ADAMS PAYS YOU A VISIT

12. EXCESSIVE DEPTH BELOW SEA LEVEL

VIDEO NOTES

The Super-Charged World of Chemistry Part 2

13. ROOMMATE'S OBSESSION WITH KENNY G

14. ROOMMATE'S OBSESSION WITH SPAM

15. ROOMMATE'S OBSESSION WITH RERUNS OF *THE MATCH GAME*

16. ROOMMATE'S OBSESSION WITH MTV'S *SPRING BREAK GRIND*

17. ROOMMATE'S OBSESSION WITH DEVELOPING A NEW BRAND OF MOUTHWASH

18. HAVING TO SHOWER WITH TOTAL STRANGERS

19. TRYING TO GET THAT HORRIBLE BURNT-TOAST SMELL OUT OF YOUR SHEETS

20. EXCITING POTENTIAL THAT YOUR LAUNDRY WILL BE STOLEN

21. BEING BROKE WHEN THE ICE CREAM TRUCK SHOWS UP

22. ROOMMATE HAS MAD COW DISEASE

23. ROOMMATE IS A MAD COW

Good news: you don't have to calculate enthalpies of formation. Since some philanthropic chemists have already worked them all out, you can just look them up on a table in your chemistry textbook. Also, the standard enthalpy of formation is zero for all elements in their stable form (like oxygen or copper), and for compounds, all you do is look up the enthalpy of formation in your textbook.

Here's another good rule to paint on your wall: **Hess's Law.** According to Hess's Law, the total enthalpy change in a reaction equals the sum of the enthalpy changes of all the intermediate steps. This means that if the total reaction consists of three reactions, you just add up the enthalpy changes of the three reactions to get the reaction's total enthalpy change.

Using Hess's Law and what we know about the standard enthalpy of formation, we can calculate a reaction's enthalpy change with this equation:

$$\Delta H^\circ_{\text{REACTION}} = \Sigma \Delta H_f^\circ{}_{\text{PRODUCTS}} - \Sigma \Delta H_f^\circ{}_{\text{REACTANTS}}$$

The equation above involves another funky Greek letter: Σ, also known as *sigma*. The wise scholars of yore used this letter to represent "the sum of" in

mathematical equations. For example, Σ 1 + 2 means "the sum of one plus two"--which works out to three. Paste a gold star on yourself for that one.

Sigma lends a hand in the equation above by shortening the mathematical statement, "the reaction's enthalpy change equals the sum of the standard enthalpy of formation of each of the products, minus the sum of the standard enthalpy of formation of each of the reactants." Basically, our equation shows that we're subtracting the stuff we started with from the stuff we end up with.

Do not make loon soup.

– Advice from *The Eskimo Cookbook (1952)*

Let's take a concrete example. We'll look at how sulfur dioxide (SO_2) and oxygen react to form sulfur trioxide (SO_3).

$$2SO_2 \ (g) + O_2 \ (g) \rightarrow 2SO_3 \ (g)$$

Where do we start? We want to find the standard enthalpy change in this reaction, so we take the standard enthalpy of formation for the product (SO_3) and multiply it by 2 to get the enthalpy (ΔH) of sulfur trioxide. The reason why we double the product's enthalpy is that there's a 2 in front of SO_3, which means we have 2 moles of it.

Next we subtract the sum of the reactants' enthalpies of formation. We multiply 2 times the standard enthalpy of formation of SO_2 (since we have 2 moles of it), then add the standard enthalpy of formation of O_2. Checking the values for the reactants' standard enthalpy of formation is a snap, since we can just look them up on the tables in any general chemistry textbook. We plug in the standard enthalpy of formation for SO_3, which is -395.2 kilojoules per mole, and the $\Sigma \Delta H_f°$ of SO_2, which is -296.9 kilojoules per mole. Oxygen's standard enthalpy of formation is zero, because the standard enthalpy of formation of any element in its most stable form is always zero.

$\Delta H_f°(SO_3) = -395.2 \text{ kJ/mol}$

$\Delta H_f°(SO_2) = -296.9 \text{ kJ/mol}$

$\Delta H_f°(O_2) = 0$

Now that we have assembled all the values we need, we can plug them in and solve the equation.

VIDEO NOTES

The Super-Charged World of Chemistry Part 2

$$\Delta H^\circ = 2\Delta H_f^\circ(SO_3) - [2\Delta H_f^\circ(SO_2) + \Delta H_f^\circ(O_2)]$$

CHANGE IN ENTHALPY = (ENTHALPY OF THE SUM OF THE PRODUCTS) – (ENTHALPY OF THE SUM OF THE REACTANTS)

$$\Delta H^\circ = 2(-395.2) - [2(-296.9) + 0]$$
$$\Delta H^\circ = -790.4 - (-593.8) = -196.6 \text{ kJ}$$

Finishing up the math, we get -196.6 kilojoules for the reaction's ΔH°. Since the change in energy is negative, we know that heat is exiting the reaction: it's an exothermic reaction.

Mole of Knowledge:

- •ENTHALPY IS THE HEAT CHANGE THAT OCCURS DURING A REACTION.

- •WHEN HEAT EXITS A REACTION, THE REACTION IS EXOTHERMIC.

- •WHEN HEAT FLOWS INTO A REACTION FROM ITS SURROUND-INGS, THE REACTION IS ENDOTHERMIC.

- •THE STANDARD ENTHALPY OF FORMATION IS THE ENTHALPY CHANGE IN A REACTION THAT FORMS A COMPOUND FROM ELEMENTS IN THEIR STANDARD STATES AT 25°C AND 1 ATMOS-PHERE OF PRESSURE.

- •HESS'S LAW SAYS THAT THE TOTAL ENTHALPY CHANGE IN A REACTION IS THE SUM OF THE ENTHALPY CHANGES OF THE REACTIONS THAT MAKE UP THE TOTAL REACTION.

V 2

106

`0:24:25` ## SECTION B: CALORIMETRY

Calorimetry measures the heat absorbed or created in chemical reactions.

`0:24:42`
`0:24:48`
`0:25:00`

The heat absorbed or created in a reaction is determined by two things: **heat capacity**, which is the amount of heat you need to raise the temperature of an object or substance by 1 Kelvin, and **molar heat capacity**, which is the amount of heat required to raise the temperature of *one mole* of a substance by 1 Kelvin. The heat capacity of *one gram* of a substance, which is the heat required to raise the temperature by 1 Kelvin, is called the **specific heat**. There should be a chart of specific heat values located conveniently in your textbook.

VIDEO NOTES

The Super-Charged World of Chemistry Part 2

Let's take a closer look at a problem you might get in class.

HOW MANY JOULES OF HEAT ARE REQUIRED TO RAISE THE TEMPERATURE OF 291 GRAMS OF LEAD FROM 21.9 DEGREES CELSIUS TO 40.8 DEGREES CELSIUS?

We make a beeline for the specific heat chart in the back of our favorite chemistry book, and discover that the specific heat of lead is 0.129 joules per gram per Kelvin. This means it takes 0.129 joules of heat to raise one gram of lead by one Kelvin. But how many joules of heat are required to raise the temperature of 291 grams of lead from 21.9 degrees celsius to 40.8 degrees Celsius?

To take this problem to the next step, we'll have to use a new formula: the quantity of heat required equals the change in temperature, multiplied by the mass, multiplied by the specific heat. This formula determines how much heat you must add to a substance to heat it up a specific number of degrees. (You'll find this formula on the insert card that came with your *Super-Charged World of Chemistry Part 2 Video Course Review.*)

$$q \text{ (heat)} = \Delta t \text{ (temperature change)} \times \text{mass} \times \text{specific heat}$$

V2

STUDY SIDEKICK

108

The problem is instantly easier, now that we have all this information ready. We have what's given in the question and we looked up the specific heat of lead (0.129 joules per grams per Kelvin), so now we just need to plug in the values.

The temperature increases to 40.8 degrees from 21.9 degrees, so we subtract and get 18.9 degrees. After plugging the heat change into the equation, we multiply by the mass we're dealing with (291 grams), and the specific heat of lead (0.129 joules over grams times Kelvin), which we found on the chart. The equation looks like this: q equals 18.9 Kelvin times 291 grams times 0.129 joules per gram per Kelvin. Dabble with the math for but a moment, and presto! We can now say with confidence that we need 709 joules of heat to raise 291 grams of lead by 18.9 Kelvin.

$$\Delta t = 40.8 - 21.9 = 18.9$$

$$mass = 291 \text{ g}$$

$$\text{specific heat of Pb} = \frac{0.129 \text{ J}}{g \cdot K}$$

$$q = 18.9K \times 291g \times \frac{0.129J}{g \cdot K} = 709 \text{ J}$$

Atomic
Advice

- Calorimetry measures the amount of heat absorbed or created in reactions.

- Heat capacity is the amount of heat required to raise the temperature of an object or substance by one Kelvin.

- Molar heat capacity is the amount of heat required to raise the temperature of *one mole* of a substance by one Kelvin.

- Specific heat is the amount of heat required to raise the temperature of *one gram* of a substance by one Kelvin.

- The specific heat and molar heat of substances are listed conveniently in your chemistry textbook.

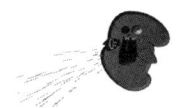

QUIZ 4

(ANSWERS ON PAGE 303)

1. Calculate the amount of energy required to raise the temperature of a 13.5 g piece of copper from 25.0 °C to 423 °C. The specific heat of copper is 0.385 J/(g·K). (Hint: Since a 1°C change equals a 1°K change, the change in temperature is the same on both scales.)

2. Calculate the total $\Delta H°$ for the following reaction. (Check your textbook for the $\Delta H_f°$ values you need.)

   $$C_2H_6 (g) + \frac{7}{2} O_2 (g) \rightarrow 2CO_2 (g) + 3H_2O (l)$$

3. Calculate the $\Delta H_f°$ for $CaSiO_3$ in the reaction below:

   $$CaO (s) + SiO_2 (s) \rightarrow CaSiO_3 (s)$$
   $$\Delta H = -89.5 \text{ kJ}$$

The Super-Charged World of Chemistry Part 2

ATOMIC STRUCTURE

So far, most of the material you've encountered in chemistry has dealt with formulas, calculations, and terms. Quantum mechanics is a little different. You won't have to calculate or measure anything when you're studying quantum mechanics, since that would involve advanced calculus. Rather, you'll need to get a grasp on the *theory* of quantum mechanics and how it's explained with numbers, letters, and pictures. Quantum mechanics helps us understand where the electrons are in atoms and molecules. It's important for us to know where the electrons are because electronic properties, like the properties that determine molecules' shapes, are directly related to the position of the electrons.

`0:28:32`

SECTION A: QUANTUM MECHANICS

`0:29:18`

Quantum mechanics is the theory of the behavior of electrons in atoms and molecules.

V2

112

Studying the behavior of electrons is a hairy process that often verges on paradox. Electrons are unbelievably tiny, so figuring out both where they are and what they're doing at any particular moment is really hard. One highly debated controversy scientists have run into is whether the electron should be considered a particle localized in space, or a wave spread out in space. This dilemma is known as **wave-particle duality.** Some scientists, like Erwin Schrödinger, dedicated their lives to studying this perplexing problem.

Werner Heisenberg was not only one of the youngest scientists ever to receive a Nobel Prize (he was 32 at the time), he also declared that it's impossible for us to determine *both* the location and momentum of an electron *at the same time.* This idea is called Heisenberg's Uncertainty Principle.

Schrödinger decided that the best description of electrons' behavior combines the particle *and* wave ideas, and most scientists agree with him. Schrödinger used complex math to construct diagrams of the possible locations of electrons in atoms and molecules. These theoretical representations, called **orbitals**, show you where the electron is 90 percent of the time.

`0:30:09`

An electron's orbital describes three things:

- the average distance between the electron and the nucleus of the atom
- the shape of the electron's **distribution** (the electron's possible location)
- the distribution's orientation in space (where it is in relation to the atom's nucleus and the other orbitals)

These three characteristics are described by three interrelated numbers called quantum numbers. The first, or principle quantum number, is represented by a lowercase n and refers to the average distance between the electron and the nucleus of the atom. The second quantum number is represented by a lowercase l, and is called the azimuthal quantum number. It describes the shape formed by the electron's distribution (the shape of the possible loca-

`0:31:26`

tion of the electron). Finally, m_l, the magnetic quantum number, describes where the orbital is in relation to the nucleus and the other orbitals.

The principle quantum number, n, refers to how far the electron is (on average) from the nucleus of the atom. n can have whole number values of 1 or greater: $n=1$, $n=2$, $n=3$, and so on. The greater the value of n, the larger the orbital, and the farther away from the nucleus the electron is more likely to range.

Azimuthal also happens to be Jinx's birthplace. Many light years ago, when atoms shared electrons freely and without regard for bonding, Jinx was born to a pair of quantum space heaters. Jinx, the subatomic love child. Jinx, who has the power to do calorimetry calculations in his head and the ability to consider enthalpy with a dispassionate, piercing gaze. Jinx, who has a knack for unclogging drains.

The Super-Charged World of Chemistry Part 2

The azimuthal quantum number, l, stands for the shape of the orbital. Schrödinger's mathematical distribution indicates the area around the nucleus in which the electron is likely to be. If you squint a little, that area may look like it's shaped like a sphere, a dumbbell, or another shape. Never fear, we'll talk more about distribution shapes and such groovin' stuff later on.

The value of l is related to the value of n in that l must be between 0 and $n-1$. For instance, if $n = 2$, then l must be 0 or 1. If $n = 3$, then l must be zero, 1, or 2; and if n is 1, then l must be 0. Four letters have been designated to stand as l's possible values: s, p, d, and f.

IF l EQUALS:	0	1	2	3
USE THE LETTER:	s	p	d	f

(IF $n = 2$ AND $l = 1$, IT'S A $2p$ ORBITAL) * * *

Substituting letters for the values of l makes it easy to express the n and l quantum numbers together. You'll also find this quantum number scheme on the insert card that came with your *Super-Charged World of Chemistry Part 2* Video Course Review.

The magnetic quantum number, m_l, describes the orientation of the orbital in space--that is, in relation to the nucleus and the other orbitals. The value of m_l is related to the value of the l quantum number, in that m_l consists of the integers between l and $-l$.

Okay, what does that mean in real life? If an orbital's l quantum number is 2, m_l will be $-2, -1, 0, 1$, and 2. If l is 3, then m_l will be $-3, -2, -1, 0, 1, 2, 3$. So whatever l is, m_l will be the set of values from $+l$ to $-l$.

FOR EXAMPLE,

if l = 2,

m_l =	−2,	−1,	0,	1,	2		

if l = 3,

m_l =	−3,	−2,	−1,	0,	1,	2,	3

The Super-Charged World of Chemistry Part 2

Let's look at the distribution associated with $l = 1$, which happens to be dumbbell-shaped. Say this dumbbell is the orbital of an electron, and Orbital Woman's hand is the nucleus. The n quantum number will tell you the distance between her hand and the ends of the dumbbell: that's how far away from the nucleus the electron spends most of its time. The angle of the dumbbell is like the m_l, or magnetic quantum number. It describes how the electron's path is oriented in space.

DISTANCE BETWEEN NUCLEUS AND ENDS OF DUMBELL = DISTRIBUTION (n)

NUCLEUS

DUMBELL = DISTRIBUTION (l)

ANGLE OF DUMBELL = ORIENTATION IN SPACE (m_l)

A group of orbitals with the same principle quantum number (n) make up an **electron shell**. The first shell is composed of all the electron orbitals with an n of 1. All the orbitals with an n value of 2 form the second shell; all the orbitals with an n value of 3 form the third shell, and so on.

Orbitals that have the same n and l values form a **subshell**. If you run into a couple of orbitals that each have an n value of 3 and an l value of 2, the electrons in these orbitals will congregate at the same distance from the nucleus, and the areas where the electrons are likely to be will have a similar shape.

The concept of orbitals and the Pauli Exclusion Principle are both useful to us when we're describing atoms with many orbitals. According to the Pauli Exclusion Principle, an orbital can hold a maximum of two electrons. This means there's an order to filling the orbitals. An element's electrons will fill the $1s$ orbital first, since it has the lowest energy and is nearest to the nucleus. Hydrogen, for example, has just one electron, so that electron goes in the $1s$ orbital. Helium has two electrons, so its $1s$ orbital is filled. Lithium, which has three electrons, fills its $1s$ orbital and sends its third electron to the $2s$ orbital. Boron has five electrons, so it fills the $1s$ and $2s$ orbitals with two electrons each, then sends its fifth electron to the $2p$ orbital.

The more electrons an element has, the more orbitals it will fill. This pattern is pretty logical, and all the elements stick to it. A glance at the periodic table will tell you the order in which the elements fill their atomic orbitals.

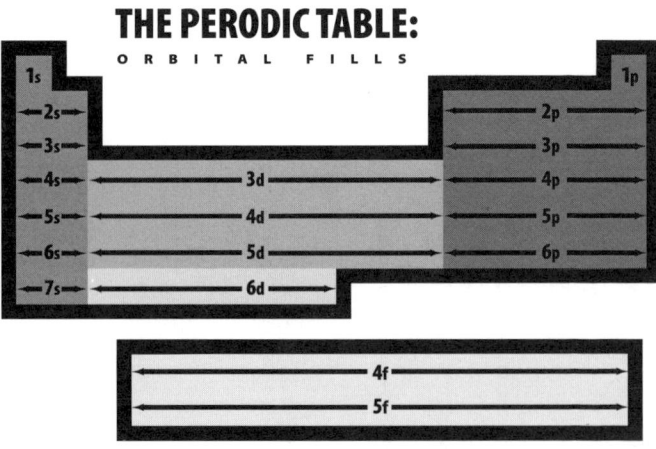

THE PERODIC TABLE:
ORBITAL FILLS

VIDEO NOTES

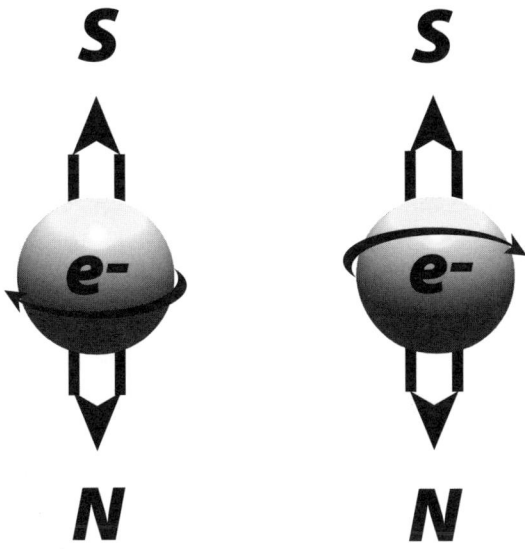

If you start at the top of the periodic table and read across from left to right, you'll see that the electrons fill more orbitals as you progress. Hydrogen and helium are labeled 1s, because they only need the 1s orbital to contain their electrons. Lithium and beryllium are both labeled 2s, because they need the 2s orbital to hold their 3 and 4 electrons, respectively. Boron, with its five electrons, needs one of the 2p orbitals. The next higher energy orbital is the 3s orbital, which both sodium and magnesium use, since sodium has 11 electrons and magnesium has 12. The next available orbital is the 3p orbital, then the 4s orbital, and so on, all the way through the 5f orbital at the bottom of the table.

Let's revisit the Pauli Exclusion Principle for a moment. Not only does this principle state that each orbital can only hold two electrons, but it also specifies that those two electrons must have opposite spins: one upward and one downward. Take helium, for example: helium has two electrons filling its *1s* orbital. One of the electrons in the *1s* orbital is what chemists call "spin-up" and the other is "spin-down." When two electrons with opposite spins fill an orbital, we consider those electrons *paired*. If an orbital has only one electron, we say that electron is *unpaired*.

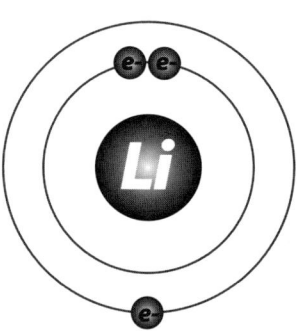

Lithium, with its three electrons, has 2 paired electrons in its *1s* orbital and one unpaired electron in its *2s* orbital. One of lithium's paired electrons is spin-up and the other is spin-down.

What do the *s*, *p*, *d*, and *f* orbitals have to do with sub-shells and quantum numbers? Let's go back and do a slow-motion replay on that orbital business.

1s orbital

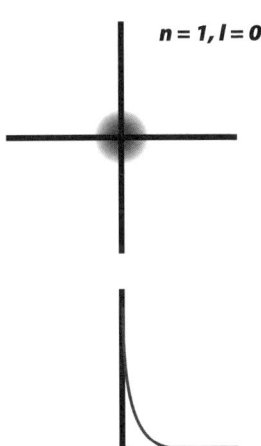

$n = 1, l = 0$

the shape of a *1s* orbital, and the dots indicate the distribution, or the path the electron may take. The area in which the dots are the densest shows where the electron is most likely to be in the orbital, statistically speaking. The intersection of the axes corresponds to the nucleus of our hypothetical atom. As you can see, *s* orbitals are sphere-shaped.

`0:38:08`

2s orbital

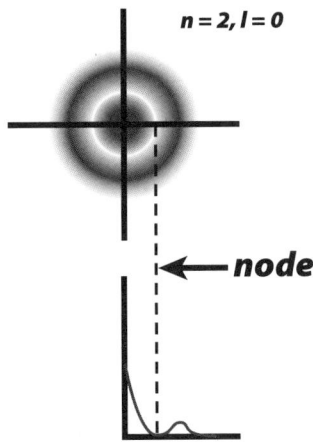

$n = 2, l = 0$

← *node*

The *1s* subshell is made up of a single orbital with an *n* of 1, an *l* of 0, and an m_l of 0. The letter *s* in *1s* stands for the *l* value, which in this case is zero, and the number 1 represents the *n* value. The diagram above shows

V2

Moving on to the 2s orbital (on the previous page), you'll notice from the density of the dots that the probable distance the electron may range from the nucleus is larger than it was in the 1s orbital. This makes sense, since the n quantum number is larger: the 2 indicates that the electron orbits farther away from the nucleus than in a 1s orbital. The diagram of the 3s orbital, which is larger yet, has two blank areas in it: these are nodes. Nodes are the places where the electron probably won't hang out, according to the math that backs up this theory.

3s orbital

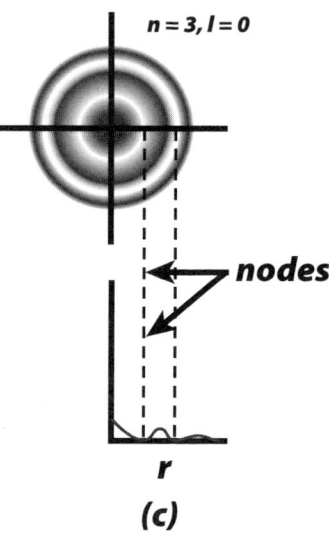

(c)

Nodes are just places in an atom in which the electrons don't travel. The number of nodes in an orbital increases as the n value increases.

The Super-Charged World of Chemistry Part 2

`0:39:29`

The $2p$ subshell contains all the $2p$ orbitals with an n equal to 2 and an l equal to 1. p orbitals are dumbbell-shaped, because the electron distribution is separated at the nucleus by a node. The node means that the electron will hang out on one side of the nucleus or the other. There are three of each of the p orbitals: three $2p$ orbitals, three $3p$ orbitals, three $4p$ orbitals, and so on. To make it easy to tell the three of each apart, we label them with the subscripts x, y, and z, which indicate the axis that the orbital is lined up with.

One essential detail to remember about the subscripts: *they don't necessarily have any relation* to the m_l quantum number that officially indicates the orientation of the orbital in space. Later on in your chemistry career, you will learn why the subscripts and the m_l quantum number don't necessarily correspond.

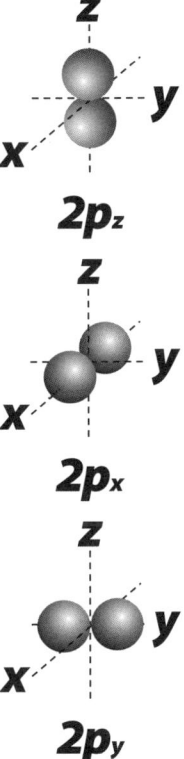

V 2

124

As we saw with the *s* orbitals, the *p* orbitals get larger as the value of *n* increases, since the larger *n* is, the farther away the electron may be from the nucleus. **Remember that the number 2 in 2p, for example, represents n.** The 4*p* electrons will be, on average, farther from the nucleus than the 2*p* electrons.

All three of the 2*p* orbitals are the same size, but they are each arranged differently in space around the nucleus. The 3*p* orbitals work the same way. Although all three of the 3*p* orbitals are the same size, they differ from each other in orientation.

I like work; it fascinates me. I can sit and look at it for hours.

– Jerome K. Jerome

The Super-Charged World of Chemistry Part 2

d orbitals appear on the scene when n is equal to 3 or more. There are always five different types of d orbitals. So there are five $3d$ orbitals, five $4d$ orbitals, five $5d$ orbitals, and so on. We use subscripts for the five different d orbitals so we can keep them straight, just like we did with the p orbitals (but again, remember that the subscripts do not always correspond with the m_l quantum numbers). Also, just like with the p orbitals, the d orbitals look similar (except for the $3d_{z^2}$ orbital) but are oriented differently.

`0:41:21`

There is one conspicuous oddball: one *d* orbital looks donut-shaped in the middle, although it has the same energy as the other *4d* orbitals.

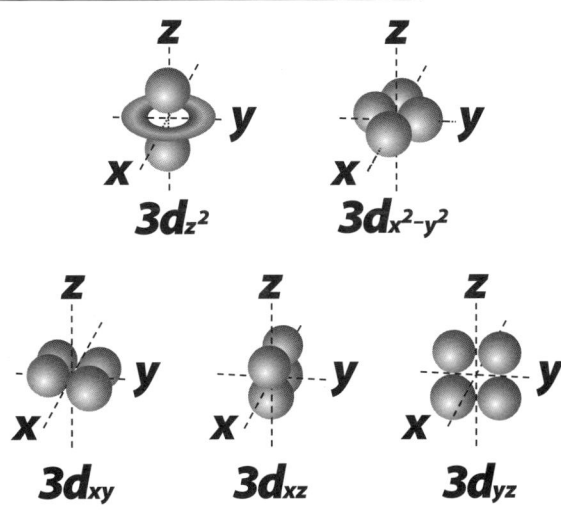

Remember that when we talk about a $3d$ orbital, for example, we are referring to an orbital with a principle quantum number of 3 and an azimuthal quantum number of 2, represented by the letter d. $4d$ orbitals have the principle quantum number 4, which means that they are bigger than the $3d$ orbitals, but they have the same azimuthal quantum number as the $3d$ orbitals, so they have the same shape.

V2

126

 0:47:11

SECTION B: WRITING ELECTRON
CONFIGURATIONS

Orbital diagrams represent **electron configurations**, or the way electrons are distributed among an atom's orbitals. Now that you know how orbitals fill--in order of increasing energy, with two electrons per orbital--you can write electron configurations.

element	total electrons	orbital diagrams 1s	2s	2p	3s	electron configuration
H	1	↑				$1s^1$
He	2	↑↓				$1s^2$
Li	3	↑↓	↑			$1s^2 2s^1$
B	5	↑↓	↑↓	↑		$1s^2 2s^2 2p^1$
Na	11	↑↓	↑↓	↑↓↑↓↑↓	↑	$1s^2 2s^2 2p^6 3s^1$

0:47:17

We'll start with hydrogen again, and diagram its electron configuration using boxes to represent orbitals and half-arrows to represent electrons. The orbital diagram for hydrogen needs just one box labeled 1s with a half-arrow pointing up. The half-arrow represents hydrogen's one electron, which occupies its 1s orbital (the lowest-energy orbital).

THE NOTATION *1s¹* ALSO REPRESENTS HYDROGEN'S ELECTRON CONFIGURATION. *1s* IS THE ORBITAL HYDROGEN USES; AND THE SUPERSCRIPT 1 TELLS US THAT HYDROGEN HAS ONE ELECTRON IN THAT ORBITAL. BREAK-DANCINGLY COOL, HEPSTERS.

Let's get cosy with another atom and see what makes it tick. Helium's two electrons are both in the $1s$ orbital, so, like hydrogen, its orbital diagram only uses one box. The difference is that helium's $1s$ box has two half-arrows that represents its two electrons. You'll notice that one half-arrow points up and one points down, showing the opposite spins of the two electrons. The notation $1s^2$ also tells us that helium fills the $1s$ orbital with 2 electrons.

Shall we attack a few more? Lithium, with its three electrons, needs two boxes for its orbital diagram. The $1s$ box is filled by a pair of electrons, and lithium's third, unpaired electron is marked in the $2s$ box. We describe lithium's electrons with this notation: $1s^22s^1$.

Boron has paired electrons filling its *1s* and *2s* orbitals, and its fifth electron remains unpaired in one of the *2p* orbitals: its notation is $1s^2 2s^2 2p^1$.

Next, we'll hop down to sodium, which has 11 electrons. Sodium fills its *1s*, *2s*, and *2p* orbitals, and sends its left-over 11th electron to the *3s* orbital. Sodium's notation is pretty lengthy, since it uses a lot of orbitals: $1s^2 2s^2 2p^6 3s^1$. Even without looking at the orbital diagrams, the notation alone tells us that sodium fills the *1s* orbital, the *2s* orbital, all three of the *2p* orbitals, and sends its unpaired electron to the *3s* orbital.

When atoms lose or gain electrons, they acquire a charge and become **ions**. The orbital diagrams for the ions reflect the addition or subtraction of the electron. Take a look at the orbital diagrams below: we've got the neutral element sodium and the positive sodium ion.

element	total electrons	orbital diagrams 1s	2s	2p	3s	electron configuration
Na	*11*	⥮	⥮	⥮⥮⥮	1	$1s^2 2s^2 2p^6 3s^1$
Na⁺	*10*	⥮	⥮	⥮⥮⥮		$1s^2 2s^2 2p^6$

The positive sodium ion has lost an electron, so its configuration is $1s^2 2s^2 2p^6$.

When an atom loses an electron, it always loses the electron from its outermost orbital. As you can see above, sodium lost the electron in its $3s$ orbital when it became an ion, because the $3s$ orbital was its outermost energy orbital.

Atomic Advice

- Orbitals are just pictures of where electrons are and what they're doing in an atom.

- Electrons will fill orbitals in a fixed pattern, starting with the lower energy orbitals closest to the nucleus, and moving up and out from there.

- You need to know where the electrons are in an atom, because the positions of the electrons affect how the atoms bond together to form molecules.

QUIZ 5

(ANSWERS ON PAGE 304)

1. Are the outermost electrons of N or P farther away from their respective nuclei? Why?

_____

2. What are the quantum numbers for the outermost electron of Na?

_____

3. Which of the following sets of quantum numbers is possible?

	n	l	m_l
a.	2	2	2
b.	1	0	0
c.	4	3	3
d.	3	1	1
e.	2	0	1

VIDEO NOTES

The Super-Charged World of Chemistry Part 2

CHEMICAL BONDING

SECTION A: LEWIS STRUCTURES

`0:51:58`

Lewis structures, which represent molecules, are named after Gilbert Lewis. Lewis developed this method of representing the covalent bonds in real molecules. Lewis structures, which may also be called electron dot structures, use dots to show the number of electrons in the outer shell (known as the *valence shell* of an atom). Lewis structures indicate the bonding pattern between the atoms that form a compound.

> *Valence is just a fancy way of saying outer.*

V2

132

`0:53:18`

WHEN WE DRAW THE LEWIS STRUCTURE OF AN ATOM, WE HAVE TO CONSIDER THE **OCTET RULE**. ACCORDING TO THE OCTET RULE, WHEN A COMPOUND FORMS, ITS ATOMS GAIN OR LOSE ELECTRONS, OR SHARE PAIRS OF ELECTRONS, UNTIL THERE ARE EIGHT ELECTRONS IN EACH ATOM'S VALENCE SHELL. ATOMS WILL KEEP SHARING UNTIL THEY HAVE 8 ELECTRONS IN THEIR OUTER SHELL. AN ATOM WITH EIGHT ELECTRONS IN ITS VALENCE SHELL IS A HAPPY, SATISFIED ATOM.

$$Na\cdot + \cdot \overset{\cdot \cdot}{Cl}:$$

$$Na \overset{\cdot \cdot}{\mp} \cdot \overset{\cdot \cdot}{Cl}:$$

$$Na^+ + \cdot \overset{\cdot \cdot}{\underset{\cdot \cdot}{Cl}}:^-$$

Here's one way the atoms in sodium and chlorine can become happy atoms. As sodium and chlorine mix, the chlorine atom removes one of the electrons from the sodium atom, leaving the sodium with a positive charge and a filled valence shell. The chlorine atom now has a negative charge and a filled valence shell. Now that's a happy looking atom! See 'em giggle and blush!

Blow in its ear.

– Johnny Carson on the best
way to thaw a turkey

The Super-Charged World of Chemistry Part 2

Don't bug out about drawing Lewis structures of simple mole-cules--there's probably only one way to draw the structure. However, there may be more than one way to draw Lewis struc-tures for more complex and beastly molecules.

Sometimes one single Lewis structure cannot describe the elec-trons in a molecule. In some cases, several versions of a molecule or ion's Lewis structure are all correct. These varying (but correct) ways of representing an atom or molecule are called **resonance structures**.

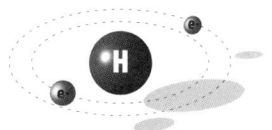

These resonance structures are all valid representations of the nitrate ion. Notice that the only difference between the represen-tations is the placement of the electrons. The same atoms must be bonded to one another in all structures; only the placement of the bond may vary. Unfortunately, there are a few exceptions to the octet rule, the most obvious exception being hydrogen.

HYDROGEN'S SHELL IS FILLED WHEN IT HAS TWO ELECTRONS; SO YOU CAN SEE THAT THERE'S NO ROOM FOR EIGHT ELEC-TRONS IN HYDROGEN'S OUTER SHELL.

V2

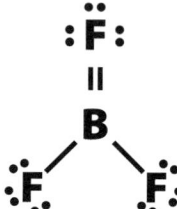

Boron trifluoride (BF_3) is another exception. There are 3 electrons in boron's valence shell and 7 in fluorine's, so 3 fluorine atoms bind to the boron atom. The molecule is complete although boron has only 6 electrons in its outer shell: it's an **electron deficient** element. (Of course, we can use resonance structures to satisfy boron's octet, like we did for NO_3^- above.)

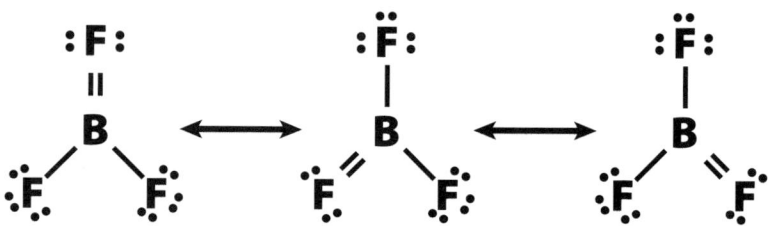

When the number of electrons in the valence shell exceeds 8, the atom has an expanded valence. This can only occur when $n \geq 3$.

`0:57:08` **SECTION B: ATOMIC BONDING**

If I Only Had an Electron
To Fill My Valence Shell

I'm missin' an electron.

It really isn't that fun. In fact, it's quite a bore.

A full valence I'd be havin', I could be a rockin' atom,

With one electron more.

`0:57:18`

There are two types of bonds that form between atoms: ionic bonds and covalent bonds.

Ions are like atoms, except ions have an electrical charge and atoms are neutral. When an electron is removed from an atom, you get a positive ion. When an electron is added to an atom, you have a negative ion.

Friendly
Reminder:
An electron has a negative charge.

Ionic bonds are the bonds that form between oppositely charged ions as they gain or lose an electron in an attempt to fill their valence shells with eight electrons. All ions and atoms want eight electrons in their valence shell.

LET'S BOOGIE ON BACK TO OUR EARLIER EXAMPLE OF THE REACTION BETWEEN SODIUM AND CHLORINE TO FORM SODIUM CHLORIDE. THE CHLORINE REMOVES AN ELECTRON FROM THE SODIUM, TURNING THE TWO ATOMS INTO OPPOSITELY CHARGED IONS. THE IONS ARE ATTRACTED TO EACH OTHER AND TO THE OTHER IONS WHICH FORM THE SODIUM CHLORIDE CRYSTAL. THAT'S A WHOLE LOT OF IONIC BONDING.

$$Na\cdot + \cdot \ddot{\underset{..}{Cl}}:$$

$$Na\overset{*}{\rightarrow} \cdot \ddot{\underset{..}{Cl}}:$$

$$Na^+ + \cdot \ddot{\underset{..}{Cl}}:^-$$

An ion's charge depends on how many electrons have been gained or lost from its valence shell. Just like atoms, an ion looking to fill its valence shell will easily combine with another ion or atom that also needs to fill its valence shell. The two ions will share electrons, allowing the ions to become happy, well-adjusted atoms with full valence shells. When ions combine to share electrons, the bond that forms between them is called a **covalent bond**.

`0:59:23`

Why not take **theRISK** & become **a happy atom?**

Meet Randolph the Electron...

· Combines easily with ion/atom that's also looking to fill valence shell.

· Shares electrons.

· Excellent cook (great pastries).

· Dances salsa, merengue, jitterbug.

VIDEO NOTES

The Super-Charged World of Chemistry Part 2

Covalent bonds usually form between non-metallic elements. Water, oxygen, and hydrogen molecules are examples of molecules with atoms that share electrons through covalent bonds.

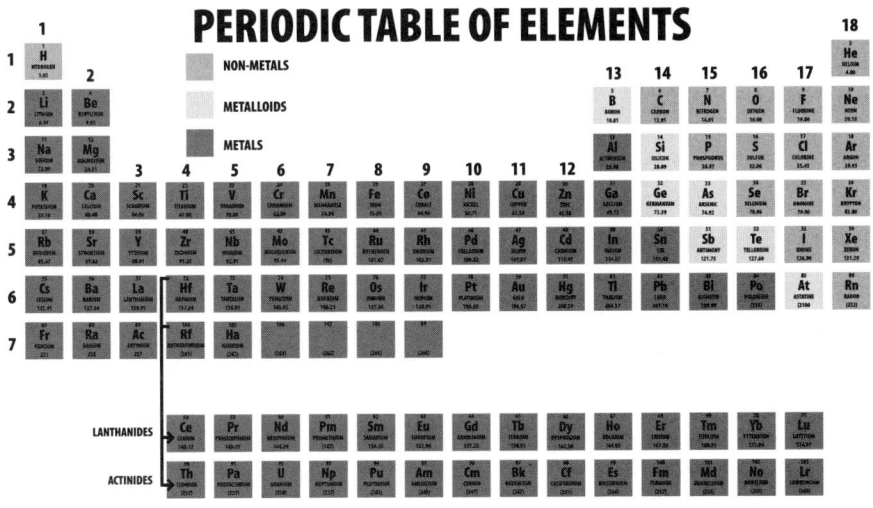

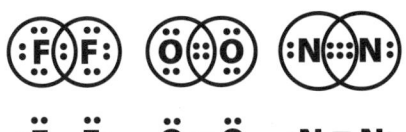

(See Stuff N⁰. 10 for a full-size periodic table!)

The Lewis structure of fluorine is a good illustration of a single covalent bond, since one pair of its electrons is shared between two fluorine atoms. Oxygen atoms share two pairs of electrons, forming a double covalent bond. Nitrogen atoms share three pairs of electrons, forming a triple covalent bond.

Electronegativity is the ability of an atom in a molecule to attract electrons bonded by covalent bonds. Some atoms are more electronegative than others, which means that in their covalent bonds with other atoms, they pull the shared electrons closer to themselves and a little farther away from the atoms they are bonding with.

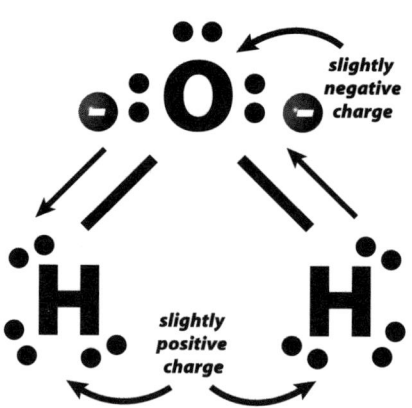

slightly negative charge

slightly positive charge

Polar covalent bonds are covalent bonds between two atoms with very different electronegativities. Let's say there's a bond between Atom X and Atom Y. Since Atom X has greater electronegativity, it pulls the negatively charged electrons away from Atom Y and closer to itself: naturally, Atom X becomes more negative. Atom Y, whose electrons are being pulled away from it a bit, assumes a weak positive charge.

We'll demonstrate this principle with a moose, a moose charmer, and a molecule of water. First we'll look at the water molecule. The oxygen atom of the water molecule has greater electronegativity than the hydrogen atoms do. As a result, the electrons of the covalent bond between oxygen and hydrogen are drawn closer to the oxygen atom, which gives a slight negative charge to the oxygen atom and a slight positive charge to the hydrogen atom.

VIDEO NOTES

The Super-Charged World of Chemistry Part 2

When a highly electronegative moose charmer does her thing, the moose can't help being drawn out of its lair inside the top hat, and closer to the moose charmer. When this occurs, the moose charmer gains in negative moose charge, and the moose's lair loses a little bit of mooseness.

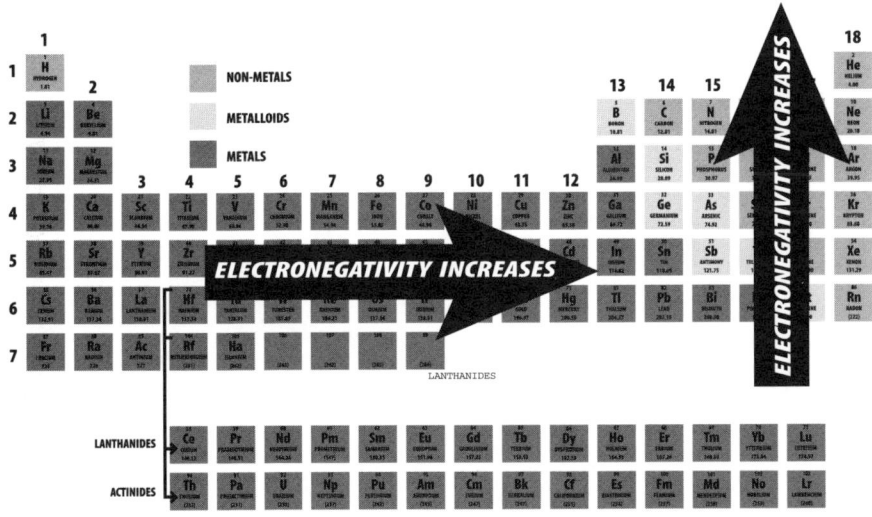

ELECTRONEGATIVITY INCREASES AS YOU GO UP AND ACROSS TO THE RIGHT OF THE PERIODIC TABLE.

Mole of Knowledge:

- IONIC BONDS FORM BETWEEN TWO OPPOSITELY CHARGED IONS.

- COVALENT BONDS FORM WHEN TWO ATOMS SHARE AN ELECTRON.

- STRONGLY ELECTRONEGATIVE ATOMS PULL THE SHARED ELEC-TRONS IN A COVALENT BOND TOWARD THEMSELVES, CAUSING THE MOLECULE TO BECOME POLAR.

- IN A POLAR MOLECULE, THE LESS ELECTRONEGATIVE ATOM HAS A SLIGHT POSITIVE CHARGE, AND THE MORE ELECTRONEGATIVE ATOM HAS A SLIGHT NEGATIVE CHARGE.

A gentleman is a man who can play
the accordion but doesn't.

– Unknown

VIDEO NOTES

The Super-Charged World of Chemistry Part 2

SECTION C: BOND ENERGY

Bond energy (BE) is the energy needed to break the bond(s) between the atoms in one mole of a substance. Say you have one mole of a gaseous substance, and each of the molecules consists of two atoms. The bond energy is the amount of energy needed to break up one mole of the molecules into single atoms.

The stronger the bonds are in a molecule, the more energy you'll need to break up the molecule into its constituent atoms. As the number of bonds between atoms increases, the bond energy also increases. That makes sense, right? More bonds, more bond energy. (Here's a perk: you can look up bond energy values in your chemistry textbook.)

You can use bond energy to estimate how much heat will be gained or lost during a reaction. The heat gained or lost during a reaction is the **reaction enthalpy**. (If you need a quick refresher on enthalpy, just turn back to Section A.)

• Δ represents a change: the final value minus the initial value

• Remember, we don't speak in absolute quantities with enthalpy, we talk about a change from some reference state.

Reaction enthalpy is the sum of the bond energies of the bonds *broken* during a reaction, minus the sum of the bond energies of the bonds *formed* in a reaction.

$$\Delta H = \Sigma BE \text{ bonds broken} - \Sigma BE \text{ bonds formed}$$

Here's a peachy reaction for you: chlorine gas plus methane gas makes chloromethane gas plus hydrogen chloride.

$$Cl_2(g) + CH_4(g) \rightarrow CH_3Cl(g) + HCl(g)$$

In this reaction, one chlorine–chlorine bond breaks, and one hydrogen–carbon bond breaks; one chlorine–carbon bond forms and one hydrogen–chlorine bond forms.

$$\Delta H = \text{energy}(Cl-Cl + H-C) - \text{energy }(Cl-C + H-Cl)$$

The Super-Charged World of Chemistry Part 2

If you'll kindly fix your X-ray eyes on the reaction enthalpy of the previous equation, it will be clear that ΔH equals the sum of the bond energies of the bonds broken minus the sum of the bond energies of the bonds formed...In this case, ΔH equals the sum of the energies of the bonds that break between a chlorine and a carbon and between a hydrogen and a carbon, minus the sum of the energies of the bonds that form between a chlorine and a carbon and a hydrogen and a chlorine.

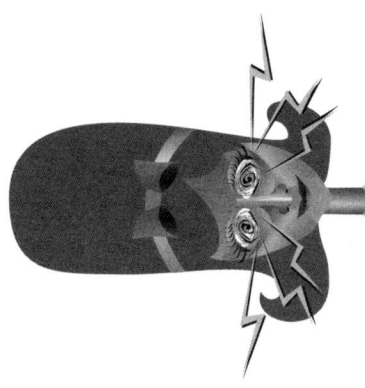

Flip nonchalantly through your textbook, find the bond energy that's given off when two chlorine atoms break apart (242 kilojoules), and add the energy given off when you break the bond between hydrogen and carbon (413 kilojoules). From that result, you casually subtract the sum of the energy given off when carbon and chlorine break apart (328 kilojoules), plus the energy given off when hydrogen and chlorine break apart (431 kilojoules).

$$\Delta H = \Sigma BE_{\text{BONDS BROKEN}} - \Sigma BE_{\text{BONDS FORMED}}$$

$$\Delta H = \text{energy}(Cl-Cl + H-C)$$
$$\quad - \text{energy}(Cl-C + H-Cl)$$

$$\Delta H = (242kJ + 413kJ)$$
$$\quad - (328kJ + 431kJ)$$

$$\Delta H = 655kJ - 759kJ$$

$$\Delta H = -104kJ$$

After some mathematical fiddling, you get the bond energy of this reaction: −104 kilojoules. Natch.

QUIZ 6

(ANSWERS ON PAGE 305)

1. Draw the Lewis structures for the following molecules or ions.

    a. NF_3 b. CCl_4

    c. NH_4^+ d. Br_2

    e. N_2 f. AlF_3

    g. NO^+

The Super-Charged World of Chemistry Part 2

2. Define the following terms or concepts.

    a. Electronegativity   _____

    b. Polar covalent bonds   _____

    c. Ionic bonds   _____

    d. Covalent bonds   _____

    e. Bond energy   _____

3. Determine the reaction enthalpy for the following reactions. (Bond energies can be found in your textbook.)

    a. $C_2H_4(g) + Br_2(g) \rightarrow C_2H_3Br(g) + HBr(g)$

    _____

    b. $4Cl_2(g) + CH_4(g) \rightarrow CCl_4(g) + 4HCl(g)$

    _____

    c. $C_2H_6(g) + 7/2O_2(g) \rightarrow 2CO_2(g) + 3H_2O(g)$

    _____

CHEM SAFETY GUY says,
HEADS UP!

Check out Practice Exams 2 and 3 in the Test Yourself section for a little brain ticklin'.

OUR EXAMS ARE 100% FIBER AND VITAMIN B-12 RICH.

VIDEO TIME CODE

The Super-Charged World of Chemistry Part 3

V3

148

VIDEO NOTES

MOLECULAR GEOMETRY AND BONDING THEORIES

Like fruit baskets and bouquets of flowers, molecules come in a variety of different shapes, and their constituent atoms may be arranged in all sorts of ways. **Molecular geometry** is the geometric arrangement of the atoms in molecules and covalently bonded crystals. You've probably seen those funky models of molecules that look like tinkertoys, right? Those are models designed to help you understand molecular geometry.

`0:02:45`

MTV is the lava lamp of the 1980's.

– Doug Ferrari

AS YOU MIGHT EXPECT, IONIC AND COVALENT BONDS EACH FORM DIFFERENT STRUCTURES. IONS ATTRACT AS MANY IONS OF OPPOSITE CHARGE AS THEY CAN PACK AROUND THEM. COVALENT BONDS, HOWEVER, HAVE A DIFFERENT METHOD. TWO COVALENTLY BONDED ATOMS SHARE A PAIR OF ELECTRONS IN A NEAT AND SPECIFIC MANNER; THEY REFUSE TO JUST CRAM TOGETHER LIKE A MESS OF SARDINES. BECAUSE COVALENTLY BONDED ATOMS ARE ARRANGED SYSTEMATICALLY, THE BOND BETWEEN THEM IS CALLED A DIRECTIONAL BOND.

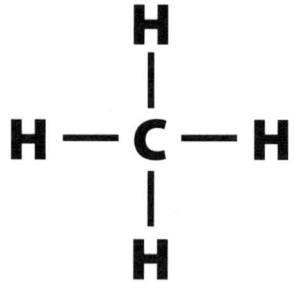

COVALENT BOND

VIDEO NOTES

The Super-Charged World of Chemistry Part 3

SECTION A: THE VSEPR THEORY

`0:03:47`

How's a cowboy to know the shape of a molecule? Scientists have developed a number of theories regarding the arrangement of atoms in a molecule. A common model used to predict molecular geometry is the **V**alence **S**hell **E**lectron **P**air **R**epulsion model, known as the **VSEPR model.** The VSEPR model describes how electron pairs bond, and how those pairs and bonds will affect the shape of the molecule. The VSEPR model theorizes that the electron pairs in a molecule's valence shell will arrange themselves as far away from each other as possible. In other words, the electrons act like they find each other repellent. These buggers have attitude and like their space.

Git off my land you wannabe cowrustler.

Go brand yourself, fool.

According to the VSEPR model, if a molecule has two pairs of shared electrons, the molecule will have a linear shape. This linear-shaped molecule is represented by the notation AX_2. The subscript 2 indicates that there are two pairs of shared electrons.

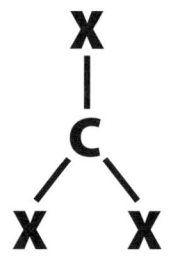

A molecule with 3 pairs of shared electrons is called trigonal and is represented by the notation AX_3. Logically enough, the subscript 3 indicates that the molecule has three pairs of shared electrons.

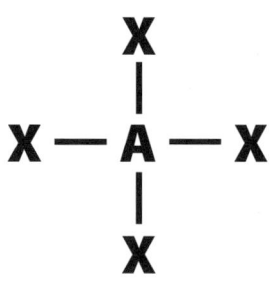

Got the pattern? How about molecules with four pairs of shared electrons? "They have a tetrahedral shape and are represented by AX_4," you say wisely.

Molecules can also have nonbonding electron pairs. The nonbonding electron pair takes the place that a regular bond would occupy.

IF YOU HUNGER FOR OTHER EXAMPLES OF THIS PHENOMENON, CHECK YOUR TEXTBOOK. OR YOU CAN JUST SPREAD A LITTLE MUSTARD ON YOUR TEXTBOOK AND GNAW ON IT.

Hunker down and take a look at this example--one of the many possible arrangements of bonds and atoms.

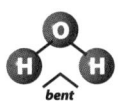

A molecule of water has two bonds between hydrogen and oxygen. The molecules' oxygen atom also has two nonbonding electron pairs. This gives water a tetrahedral arrangement of electron pairs and bonds. But since diagrams of molecules don't show the nonbonding electrons of oxygen, we just say that the water molecule is *bent* rather than tetrahedral.

Knowing the geometric arrangement of the bonds in a molecule comes in handy when you want to determine a molecule's polarity. The only other thing you need to figure out polarity is the **electronegativity** of the atoms, or how strongly some of the atoms pull at the shared electrons. That is, you need to be up on which atoms are electron hogs.

As we discussed in the section on chemical bonding in *The Super-Charged World of Chemistry Part 2*, if a molecule is made up of two atoms with different electronegativities, the molecule is held together by polar covalent bonds. The atom with greater electronegativity is by nature piggier with its electrons, so it holds onto them more tightly. The atom thus acquires a slight negative charge for itself, and the other

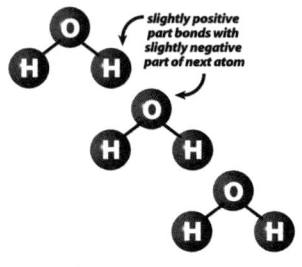

slightly positive part bonds with slightly negative part of next atom

atom (the one that gets left out in the rain) gains a slight positive charge. This kind of molecular arrangement is called a **dipole**.

Although dipole molecules have slightly positive and slightly negative regions in them, they are still considered neutral molecules. The small charges in certain regions of the molecule just affect the way a group of these molecules would line up alongside each other. Dipole molecules tend to line up so that the negative part of one molecule pairs with the positive part of the next. These attractions between polar molecules are known as **dipole-dipole forces**.

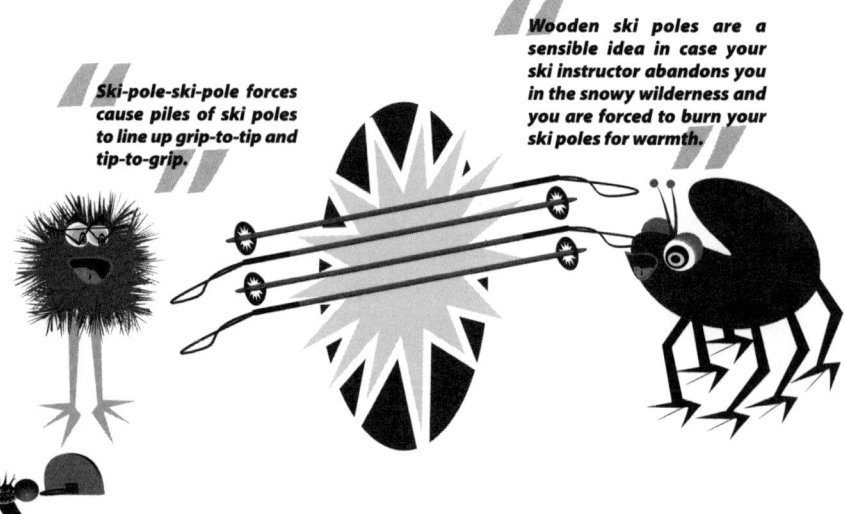

Ski-pole-ski-pole forces cause piles of ski poles to line up grip-to-tip and tip-to-grip.

Wooden ski poles are a sensible idea in case your ski instructor abandons you in the snowy wilderness and you are forced to burn your ski poles for warmth.

The Super-Charged World of Chemistry Part 3

Remember in *The Super-Charged World of Chemistry Part 2* we said that you needed to know where the electrons are in molecules and what they're up to? Now you know why! The VSEPR theory counts on the fact that the distribution of electrons affects both the shape and behavior of molecules. Water molecules, for example, stick together because they are polar molecules. It all comes back to the placement of those electrons!

`0:06:45`

 SECTION B: ORBITAL OVERLAP

So how do atoms bond together to form a molecule? There are basically two accepted theories that address this question: the hybrid orbital theory and the molecular orbital theory.

 Hybrid Orbital Theory

When molecules form, the orbitals of the component atoms overlap to produce **hybrid orbitals**. These hybrid (overlapping) mixtures of the atomic orbitals allow two electrons with opposite spins to come together to form a covalent bond, which thereby helps hold the molecule together. Like all atomic orbitals, each hybrid orbital can only hold two electrons.

Some Failed EXPERIMENTS
in Creating Hybrids:

THE AQUA-BIKE

THE CHICK TERRIER

BORK! BORK! BORK!

THE STANANA

THE ABOMINABLE SNOW CAMEL

THE GRAHAM CRACKER SOLAR PANEL

A molecule's Lewis structure shows you how many hybrid orbitals the molecule will need for it to accommodate all its bonds and unbonded electron pairs. Some of the atoms' valence orbitals will combine to form the hybrid orbitals.

You need a hybrid orbital for each bonded and unbonded pair of electrons in a molecule. For example, in beryllium hydride, beryllium is the central atom and the hydrogens are bonded to it. According to the molecule's Lewis structure (shown below), two bonds must be accommodated by hybrid orbitals, so BeH_2 has two hybrid orbitals.

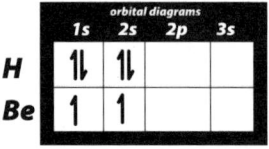

Beryllium's $1s$ and $2s$ orbitals are filled in its orbital diagram. Hydrogen's orbital diagram shows that it only has a single electron in the $1s$ orbital. Since we have two hydrogens, each with a single electron in its $1s$ orbital, we want to set up the diagram of BeH_2 so that each of the hydrogens' electrons bonds with one of beryllium's electrons, and each bonded pair gets a hybrid orbital of its own.

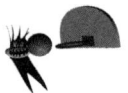

The Super-Charged World of Chemistry Part 3

There's a neat trick to arranging beryllium's electrons so they can bond with hydrogen's electrons. We "promote" an electron from the *2s* orbital to the *2p* orbital, so that beryllium's orbital diagram looks like this:

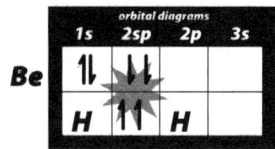

Once there is only one electron in the *2s* orbital and one electron in the *2p* orbital, beryllium's *s* and *p* orbitals mix to form two *sp* hybrid orbitals. These *sp* hybrid orbitals can bond with hydrogen's orbitals.

Hybrid orbitals will always arrange themselves as far from each other as possible, and in this case, the orbitals are 180 degrees from each other.

Conveniently for the world's chemistry students, there's a hip and speedy way to figure out how many hybrid orbitals will form in a molecule: **hybrid orbital numbers.**

You appeal to a small, select group of confused people.

– Fortune cookie

The hybrid orbital number of a molecule is the number of bonds and unbonded electron pairs that form around the atom in the center of the molecule. When you count bonds for hybrid orbital numbers, double and triple bonds count as a single bond. The chart below tells you how the orbitals will hybridize based on the hybrid orbital number.

hybribid orbital #	hybridization	geometry
2	sp	linear
3	sp^2	trigonal planar
4	sp^3	tetrahedral
5	dsp^3	trigonal bipyrmid

Let's unravel a small problem.

How do the hybrid orbitals form around boron in the compound boron trifluoride (BF₃)?

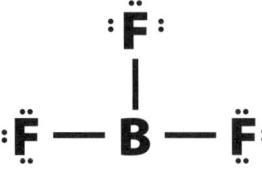

Plunge right into the Lewis structure for BF₃. Count the number of single bonds and lone electron pairs around boron (the atom in the center). There are three single bonds and no lone pairs, so the hybrid orbital number is 3. Examine the chart, and you'll see that to accommodate the three bonds, the molecule needs three sp^2 hybrid orbitals.

The Super-Charged World of Chemistry Part 3

The three bonds in BF_3 form a trigonal planar arrangement. Beautiful.

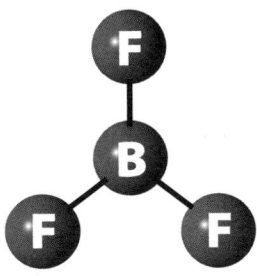

Methane (CH_4) is another kettle of fish—it forms a three-dimensional molecule. You can tell by its Lewis structure that methane has four single bonds. Check the handy chart, and you'll see that four bonds gives methane a hybrid orbital number of 4, and a three-dimensional tetrahedral arrangement.

$$
\begin{array}{c}
\mathbf{H} \\
| \\
\mathbf{H} - \mathbf{C} - \mathbf{H} \\
| \\
\mathbf{H}
\end{array}
$$

Ignore previous cookie.

– Fortune cookie

Molecular Orbital Theory

Molecular orbital theory is another model that explains how molecules manage to hang together. According to this theory, the orbitals of the atoms in the molecule overlap, forming molecular orbitals that glue the nuclei together. When two atomic orbitals overlap, they form two molecular orbitals. Hydrogen's orbitals illustrate this theory.

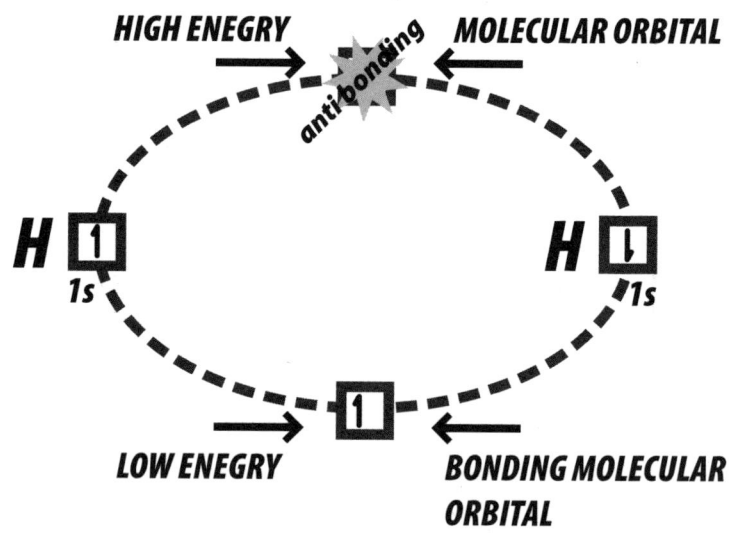

The two overlapping $1s$ hydrogen orbitals form two molecular orbitals --one with high energy and one with low energy. The low-energy **bonding molecular orbital** forms the covalent bond that holds the nuclei together. The friendly bonding electrons in the bonding molecular orbital crowd between the two hydrogen nuclei and join in the covalent bond that holds the atoms together. The high-energy **antibonding molecular orbital**, however, doesn't help hold the molecule together at all. This orbital not only has fewer electrons, but the high energy and bad attitudes of the antibonding electrons keep them from bonding. Each molecular orbital can hold two electrons with opposite spins, just like atomic orbitals.

In the world of molecular orbital theory, **bond order** refers to a way of counting bonds between two atoms in a molecule. More bonds make a molecule more stable, so a high bond order number indicates a molecule that will hold together well.

Antibonding electrons have a bad attitude.

To find bond order, subtract the number of antibonding electrons in the molecular orbitals from the number of bonding electrons, then divide that number in half. The reason you divide this number in half is that you are counting the number of *pairs* of electrons, not the number of single electrons.

Bond Order = ½ (# of bonding electrons − # of antibonding electrons)

Calculating bond order is pretty straightforward. Let's go back to the hydrogen example. Hydrogen has two bonding electrons and zero antibonding electrons, so the equation we'll use to find hydrogen's bond order looks like this.

number of bonding electrons in H_2 = 2

number of antibonding electrons in H_2 = 0

Bond order of H_2 = ½ (2−0) = 1

How do you know how many bonding electrons and antibonding electrons there are? It's not too hard, actually. Just keep in mind that each orbital can only hold two electrons, and electrons will always go to the lowest energy orbital first. For instance, say you have a molecule with five electrons. Two will go to the first bonding molecular orbital, the next two will go to the first

antibonding molecular orbital, and the fifth will go to the second bonding molecular orbital. You'll have a total of three bonding electrons and two antibonding electrons.

Let's try another one on for size. Say your molecule has to accommodate a total of eight electrons. Two will go to the first bonding molecular orbital, two will go to the first antibonding molecular orbital, two will go to the second bonding molecular orbital, and two will go to the second antibonding molecular orbital. In this case, you have four bonding electrons and four antibonding electrons.

Mole of Knowledge

- TWO MODELS EXPLAIN HOW MOLECULES HOLD TOGETHER: THE HYBRID ORBITAL THEORY AND THE MOLECULAR ORBITAL THEORY.

- ACCORDING TO THE HYBRID ORBITAL THEORY, WHEN THE ORBITALS OF TWO ATOMS IN A MOLECULE OVERLAP, THEY FORM A COVALENT BOND THAT HOLDS THE MOLECULE TOGETHER.

  - THE NUMBER OF OVERLAPPING ORBITALS DEPENDS ON HOW MANY ARE NEEDED TO ACCOMMODATE ALL THE PAIRS OF BONDED AND UNBONDED ELECTRONS.

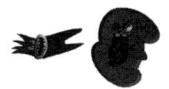

STUDY SIDEKICK

• ACCORDING TO THE MOLECULAR ORBITAL THEORY, WHEN TWO ORBITALS COME TOGETHER (ONE FROM ONE ATOM AND ONE FROM ANOTHER), TWO NEW ORBITALS WILL BE FORMED.

– ONE OF THESE TWO NEW ORBITALS WILL HAVE LOW ENERGY AND ONE WILL HAVE HIGH ENERGY.

– THE LOW ENERGY, OR BONDING MOLECULAR ORBITAL, FORMS A COVALENT BOND THAT HOLDS THE MOLECULE TOGETHER.

– THE HIGH ENERGY, OR ANTIBONDING MOLECULAR ORBITAL, DOES NOT HELP HOLD THE MOLECULE TOGETHER AT ALL; IN FACT, WHEN THE ANTIBONDING MOLECULAR ORBITAL IS FILLED, IT CANCELS OUT THE BONDING MOLECULAR ORBITAL.

Eat cheese or die.

– Joel McNally's motto for Wisconsin

VIDEO NOTES

The Super-Charged World of Chemistry Part 3

QUIZ 7

(ANSWERS ON PAGE 308)

1. Using the VSEPR model, predict the geometry of the following molecules.

    a. $BeCl_2$ _____

    b. BF_3 _____

    c. NH_3 _____

    d. H_2O _____

    e. CCl_4 _____

2. Looking back at the previous question, what is the hybridization of each compound's central atom?

3. Define the following terms.

    a. antibonding molecular orbital

    b. bond order

    c. covalent bond

V3

GASES

Quick, take a breath! Mmmm, yum. You just sucked up a bunch of good old gases, and probably a few dust mites, too. The air you breathe is made up of elements like nitrogen and oxygen, which are gases under the normal conditions of the earth's climate. Some substances, like water, can be in a liquid, solid, *or* gaseous state under ordinary conditions. For instance, in a closed container of ice water, you've got ice, which is a solid; water, which is a liquid; and water vapor, which is a gas formed by the evaporating water. You've got all three states in that one container.

VIDEO NOTES

The Super-Charged World of Chemistry Part 3

Gases act differently from solids or liquids. For one thing, they expand and compress easily. If you put a little gas in a helium tank, for example, the gaseous molecules will move away from each other to fill up the whole container.

A LITTLE GAS ADDED   *SOME MORE GAS ADDED*

On the other hand, if you really dig that helium gas and decide to pump a whole lot of it into the tank, you shouldn't have a problem, because the gaseous atoms or molecules will shove over and pack closely together to allow more atoms in.

You describe gases based on four properties: mass, volume, pressure, and temperature.

LOTS OF GAS

The length of a film should be directly related to the endurance of the human bladder.

– Alfred Hitchcock

V 3

Think of it this way. Imagine you're talking about a really kickin' CD. You might say...

- · "It's a sizzling new band." → temperature

- · "The CD sounds best when played at the decibal-level of an elephant charge." → volume

- · "The lyrics are way heavy." ("My wife left me and took my crayons, sob, now I have no choice but to become a sumo wrestler") → mass

- · "It's a laid-back collection of tunes you can mellow out to." → pressure

Use grams and kilograms to measure the mass of a gas, and liters and milliliters for volume.

The pressure of a gas is the force it exerts on a unit of area. In other words, when we measure the **pressure** of a gas, we're measuring how hard the gas is pushing on the stuff around it. As we said before, the unit used to measure pressure is the *atmosphere*. *Atmospheric pressure* is another can of worms, though --it's the force exerted by the atmosphere on a unit of the earth's surface. So don't confuse the unit *atmosphere* with the term *atmospheric pressure*.

When you're doing gas equations, use Kelvin as your unit of temperature. Your professor might ask you to calculate a problem involving gases, and craftily give you the temperature in degrees Celsius--which then needs to be converted to Kelvins. But don't be baffled. Just add 273.15 to the Celsius reading, since 0°C equals 273.15 K.

So if you're given 22°C, you just write 295.15 Kelvins swiftly and surely. And so on, for any any other Celsius temperature you encounter. You get the idea.

WHEN YOU'RE TALKING ABOUT TEMPERATURE ON THE KELVIN SCALE, YOU DON'T NEED TO SAY "DEGREES KELVIN." THE UNITS AREN'T DEGREES, THEY'RE SIMPLY CALLED *KELVINS*.

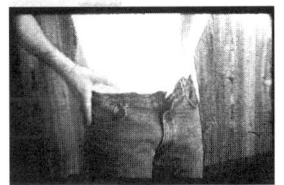

$$T(K) = t(°C) + 273.15$$

$$T(K) = 0°C + 273.15$$
$$= 273.15 \text{ Kelvins}$$

$$T(K) = 22°C + 273.15$$
$$= 295.15 \text{ Kelvins}$$

V3

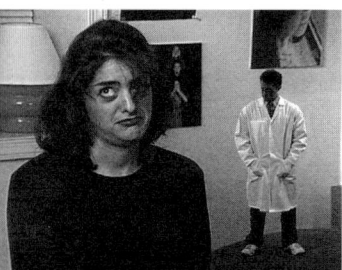

`0:21:32` SECTION A: KINETIC-MOLECULAR THEORY

Gases act differently from solids and liquids. The kinetic-molecular theory, which is based on the following five assumptions, explains why gases act the way they do.

1. A gas is composed of molecules that are far apart from each other. (Of course, we are talking about distance within a molecule, so you have to think in terms of tiny dimensions.) Most of the volume a gas occupies is nothing but empty space.

CACHE AND FRED

All these theories! Can't chemists ever just say anything for sure?

No, not in this lifetime. Even those three words can cause a lot of little problems.

How do you know that?

OH, RIGHT.

I-I-I-I hypothesize that I really like spending time with you.

HA! HA! HA!

SPACE IS SUCH A GAS!

2. Gas molecules are in constant, random motion. Each molecule continues to move in a straight line unless it collides with another molecule or with a wall of the container.

3. Gas molecules exert no force on each other or on the container, except when they collide with each other or with the walls of the container.

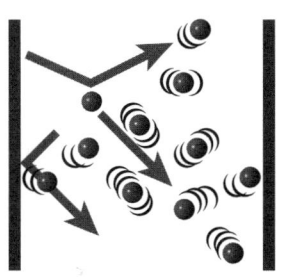

4. The average kinetic energy of the molecules in a gas is proportional to the temperature of the gas.

This one may sound complex, but think about it this way: when you heat up a pot of water, the molecules start moving really fast, and then escape from the pot (boil off). The movement of the molecules, which is the gas's kinetic energy, becomes more rapid as the temperature increases--so, the kinetic energy is proportional to the temperature.

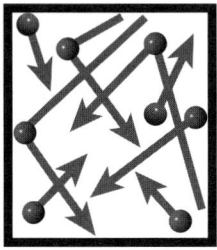

5. Every time a molecule collides with a wall, it exerts a force on the wall.

If you take assumption #5 a step further and shrink the size of the container of gases (decrease the volume), the number of molecules per unit of volume increases. If you have the same amount of gas but only half the volume, the pressure of the gas doubles; as a result, the molecules collide more frequently with the walls. The principle explaining this is called **Boyle's law**.

`0:24:04`

Although they aren't part of the five assumptions of kinetic-molecular theory, there are two more important properties of gases that you should know about: diffusion and effusion.

Diffusion is the process by which one gas mixes with another as a result of a concentration gradient.

Effusion is the process by which a gas escapes from a container through a very small hole under a pressure gradient. If you open a helium-filled balloon and let the helium escape, the helium gas effuses into the air. Once the molecules escape, we can assume that they feel, well, effusive.

Monopoly Tip: Gas molecules tend to be **generous** with even their most **valuable properties**, such as Park Place and Boardwalk--so chances are, you'll be able to talk them into **trading** early on in the **game**.

The rate of diffusion or effusion of a gas is inversely proportional to the square root of its molar mass. Translation: the heavier a gas is, the slower it diffuses or effuses. This principle is called Graham's Law.

GRAHAM'S LAW

`0:25:11`

0:25:18 **SECTION B: THE IDEAL GAS EQUATION**

The ideal gas equation is useful because it shows us how the pressure, volume, temperature, and number of moles in a hypothetical gas affect the gas's behavior. Of course, the equation is hypothetical, since in real life there are outside factors that will always interfere with gases' behavior.

Ideal Gas Equation: PV = nRT

P = pressure in atm

V = volume in liters

n = number of moles

R = gas constant

T = temperature in Kelvin

the constant $R = \dfrac{0.0821 \text{ atm-L}}{\text{mol-K}}$

The Super-Charged World of Chemistry Part 3

You can use the ideal gas equation to find the **partial pressure** of a gas. The partial pressure of a gas is the pressure exerted by one of the gases in a mixture. Dalton's law of partial pressures says that in a mixture of gases, the total pressure exerted by the gas equals the sum of the pressures that each gas would exert if it were alone under the same conditions.

DALTON'S LAW OF PARTIAL PRESSURES

`0:27:20`

$$P_{total} = P_1 + P_2 + P_3 + ...$$

Here's an application of Dalton's law. Suppose we have a one-liter flask containing 5.00 grams of carbon dioxide and 2.00 grams of oxygen molecules, at a temperature of 25 °C. What we've got to do is find the partial pressures of each of the gases and the total pressure in the flask. We can use the ideal gas equation, PV=nRT, but first we have to rearrange it so it's more useful to us--we'll set it up to find the pressure (P) since that's what we're looking for.

$$PV = nRT$$

$$P = \frac{nRT}{V}$$

In the rearranged equation, pressure equals the number of moles times the gas constant, times the temperature, divided by the volume.

The first thing to do here is convert our temperature from 25°C to Kelvins. We add 273.15 to the 25°C and get 298.15 K.

$$25°C + 273.15 = 298.15K$$

V 3

Now we have to determine how many moles of each gas are in the flask. The equation below is what we'll use for the conversion. If you watched *The Super-Charged World of Chemistry, Parts 1 and 2,* I bet you could teach your hamster how to convert from grams to moles, but here's the equation again in case either of you needs a refresher. (Flip to Section E of Stoichiometry in *The Super-Charged World of Chemistry, Part 1,* for more details.)

$$n = (\text{\# of moles}) = 1 \text{ mole substance} \times \frac{\text{wt in grams (of sample)}}{\text{wt of 1 mole in grams (of substance)}}$$

To find the partial pressure in the flask, we convert our sample from grams to moles.

$$n = 2.00 \text{ g } O_2 \times \frac{1 \text{ mol } O_2}{32.0 \text{ g } O_2}$$

$$= 0.0625 \text{ mol}$$

Remember, we rearranged the ideal gas law, $PV = nRT$, to solve for pressure.

$$P = \frac{nRT}{V}$$

$$n = 0.0625$$

$$R = 0.0821 \text{ L-atm/mol-K}$$

$$T = 298.15K$$

VIDEO NOTES

The Super-Charged World of Chemistry Part 3

The pressure in the flask equals 0.0625 moles times the gas constant. You need to choose one of the two gas constants; since you're looking for pressure, and since the total flask is measured in liters, use gas constant in terms of atmospheres and liters. That's 0.0821 liters times atmospheres per moles times Kelvin. Multiply those two quantities by the temperature (298.15 Kelvins), and divide all that by the volume, 1 liter.

Quick, faithful math-heads! Here comes a little more algebra. The denominator in the gas constant (moles times Kelvin) shifts down to the denominator of our equation. So we'll cross out all the terms in the equation and multiply the terms in the numerator. That leaves us with 1.53 atmospheres, the pressure of the oxygen in the flask.

$$P\,(O_2) = \frac{(0.0625 \text{ moles})(0.0821 \text{L-atm/mol-K})(298.15\text{K})}{1.00 \text{ L}}$$

$$= \frac{(0.0625 \text{ moles})(0.0821 \text{L-atm})(298.15\text{K})}{1.00\text{L mol-K}}$$

$$= \frac{(0.0625)(0.0821)(298.15)}{1.00} \text{ atm}$$

$$= 1.53 \text{ atm}$$

V3

Onward and upward to carbon dioxide. Here, again, we have to convert from grams to moles. We multiply one mole of carbon dioxide times the weight of our sample, 5 grams, and divide by the weight of one mole of carbon dioxide (44 grams). We now have 0.114 moles.

$$n = \frac{1 \text{ mol CO}_2}{44.0 \text{ g}} \times 5.00 \text{ g} = 0.114 \text{ moles}$$

Now we're ready to find the pressure of the carbon dioxide in the flask. The pressure equals 0.114 moles times the gas constant (0.0821 liters times atmospheres per moles times Kelvin), times the temperature (298.15), all divided by one liter. Again, we'll move the denominator of the gas constant to the denominator of the equation, and cancel out the units. We get 2.79 atmospheres, the pressure of the carbon dioxide in the flask.

$$P = \frac{nRT}{V}$$

$$P \text{ (carbon dioxide)} = \frac{(0.114 \text{ mol})(0.0821 \text{ L-atm/mol-K})(298.15 \text{ K})}{1.00 \text{ L mol-K}}$$

$$= \frac{(0.114)(0.0821 \text{atm})(298.15)}{1.00} \text{ atm}$$

$$= 2.79 \text{ atm}$$

Finally, to figure out the total pressure in the flask, just add up the individual partial pressures.

$$\text{Total Pressure } (P_T) = P_{\frac{2}{32}} + P_{C\frac{2}{3}}$$
$$= 1.53 \text{ atm} + 2.79 \text{ atm}$$
$$= 4.32 \text{ atm}$$

Zzzzap! The total pressure is 4.32 atmospheres.

Here's an equation that should look familiar--you've been using it to do conversions. The equation uses the unit factor method to change from grams to moles: you multiply the weight of your sample (in grams) by one mole of the sample, then divide that amount by the molecular or formula weight of the sample (in grams per mole).

$$\text{moles} = \frac{1 \text{ mol substance}}{\text{wt of 1 mol of substance}} \times \text{wt in grams of sample}$$

Shortcut,
shortcut!

Ready for a simplified version? You can just ignore the "1 mole of substance" in the numerator, because it doesn't do anything to change the values in your equation. Just take a shortcut and put the mass of your sample right over the molecular or formula weight of your substance, and calculate away.

$$n = \frac{1 \text{ mol of } CO_2 \times 5.00 \text{ g}}{44.0 \text{ g}}$$

$$= 0.114 \text{ moles}$$

Let's take another look at how we just converted carbon dioxide from grams to moles in our pressure calculation. Using the simplified version of the equation, we put the mass of the sample of carbon dioxide (5.00 grams) over the molecular weight of carbon dioxide (44.0 grams per mole). Cancel out the "grams" and divide 5.00 grams by 44.0 grams per mole, and you get the same answer you got before: 0.114 moles.

Anybody who thinks of going to bed before 12 o'clock is a scoundrel.

– Samuel Johnson

$$n = \frac{5.00 \text{ g}}{44.0 \text{ g/mol}}$$

$$= 0.114 \text{ moles}$$

That fancy footwork can save you some time.

When you work with the ideal gas law, keep in mind that there will be deviations from ideal behavior. In real life, gases often find an excuse not to obey the ideal gas law.

Gases have three main excuses for not obeying the law:

1. When molecules are crammed into a very small space under high pressure, the molecules take up a significant fraction of the empty space, and this can skew your results.

2. Intermolecular forces act between the molecules. (We will learn more about intermolecular forces in the next section.)

3. A rebellious streak and a free-spirited nature.*

*Your chem instructor probably won't accept this as an answer.

Mole of Knowledge

• THE IDEAL GAS EQUATION DESCRIBES THE PRESSURE, VOLUME, AND TEMPERATURE OF ANY HYPOTHETICAL GAS.

• YOU CAN REARRANGE THE IDEAL GAS EQUATION TO FIND THE PARTIAL PRESSURES OF GASES IN A CONTAINER. WHEN YOU ADD UP THE PARTIAL PRESSURES, YOU GET THE TOTAL PRESSURE OF THE GASES IN THE CONTAINER.

• THE IDEAL GAS EQUATION DESCRIBES HYPO-THETICAL GASES THAT BEHAVE EXACTLY THE WAY THEY ARE SUPPOSED TO. IN REAL LIFE, GASES DON'T ALWAYS BEHAVE LIKE THE IDEAL GAS EQUATION SAYS THEY SHOULD.

• USE THE IDEAL GAS LAW ($P = \dfrac{nRT}{V}$) TO FIND THE PRESSURE OF EACH GAS AND USE DALTON'S PARTIAL PRESSURE LAW TO FIND THE TOTAL PRESSURE.

VIDEO NOTES

The Super-Charged World of Chemistry Part 3

QUIZ 8

(ANSWERS ON PAGE 309)

1. What are the five assumptions of the kinetic-molecular theory?

2. If there are 7.0 grams of nitrogen (N_2) in a 1.00 liter container at a temperature of 25°C, what is the pressure exerted by the nitrogen?

3. If 10.0 grams of neon (Ne) are added to the container in the previous question, what is the partial pressure of Ne? What is the total pressure in the container?

188

STATES OF MATTER

`0:38:34` ## SECTION A: INTERMOLECULAR FORCES

Intermolecular forces are the forces of attraction `0:38:40` between molecules. They are not as strong as ionic and covalent bonds, but there is no doubt that they have an effect on states of matter. Since the inter-molecular forces operating on gas molecules are not that strong, they allow gas molecules to move away from each other. The intermolecular forces between molecules in a liquid are a bit stronger, causing liquids to hold together and assume the shape of whatever container they're in. Strong intermolecular forces are at work between the molecules in a solid, locking the molecules in place.

For example, water conforms to the shape of its container.

Intermolecular forces are responsible for gas molecules condensing into their liquid state. Gas molecules normally move pretty rapidly, but as the temperature decreases, the speed (kinetic energy) of the molecules also decreases. As gas molecules slow down, they begin sticking together and eventually form a liquid.

The Super-Charged World of Chemistry Part 3

Although liquids are denser than gases, molecules still have some kinetic energy when they are in their liquid state. As their temperature decreases, molecules lose more of their kinetic energy, their speed slows down, and the attractive forces between them cause them to congregate and solidify. Presto, we have another change of state: the liquid becomes a solid.

There are four types of intermolecular forces: ion-dipole interactions, dipole-dipole interactions, London dispersion forces, and hydrogen bonds.

Ion-dipole interactions are very important when a solid substance made of ions is dissolved in a liquid consisting of polar molecules. For example, when an ionic solid like sodium chloride is dissolved in a polar liquid like water, the solid salt crystals break up and form the ions Na^+ and Cl^-. The positive charge on the polar molecule moves towards the negative ion of the salt and vice versa.

`0:40:37`

Dipole is just another term for **polar molecule**. **Dipole-dipole forces**, which result from the slight charges on the ends of polar molecules, are what cause these molecules to stick together. (See the first section of *The Super-Charged World of Chemistry Part 3*.)

`0:41:12`

V3

190

 London dispersion forces are the only intermolecular forces between nonpolar atoms and molecules. London dispersion forces are inter-molecular forces that result from fluctuating, instantaneous dipoles.

ELECTROSTATIC ATTRACTIONS

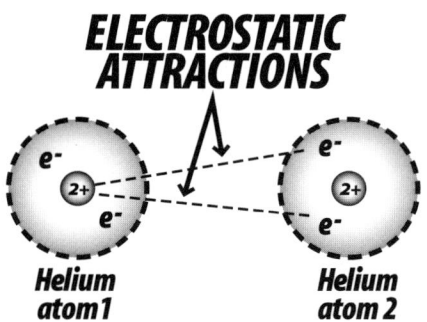

Helium atom 1   **Helium atom 2**

Ya gotta do what ya gotta do.

– Sylvester Stallone in *Rocky IV*

LET'S BREAK THAT LAST ONE DOWN. IN PLAIN ENGLISH, YOU CAN'T HAVE DIPOLE-DIPOLE FORCES BETWEEN NONPOLAR ATOMS AND MOLECULES, BECAUSE THE MOLECULES DON'T HAVE ANY CHARGES WITH WHICH TO ATTRACT EACH OTHER. HOWEVER, A GUY NAMED FRITZ LONDON FIGURED THAT SINCE NONPOLAR MOLECULES CAN REACH A LIQUID STATE, THERE MUST BE SOME INTERMOLECULAR FORCE AT WORK BETWEEN NONPOLAR MOLECULES. THAT'S HOW HE CAME UP WITH THE IDEA OF INSTANTANEOUS DIPOLE MOVEMENT.

Fritz London figured that even nonpolar atoms like helium might sometimes have both electrons on one side of the nucleus for just an instant. Ol' Fritz called this coincidence *instantaneous dipole movement*. For just an instant, a normally nonpolar atom or molecule is polar. Of course, the electrons move immediately and the atom or molecule is no longer polar.

When an atom becomes polar for an instant, the force of its electrons on one side repels the electrons of the atom next to it, shoving both electrons to one side so *that* atom is polar for a moment. This can continue in a domino effect, until a whole bunch of the atoms are polar for an instant: a temporary dipole-dipole type attraction between normally nonpolar particles. London dispersion forces operate between polar molecules, too, increasing the attraction between the polar molecules.

Hydrogen bonds are intermolecular forces that only apply to hydrogen. When a hydrogen atom bonds to a strongly electronegative atom, it forms a polar bond. The electronegative atom pulls the hydrogen's electron toward itself. Once the hydrogen's electron has been pulled away from it, hydrogen has a tiny positive charge. At this point, the hydrogen atom is attracted to an unshared electron in another atom. This attraction, found in molecules like water and hydrogen fluoride, is the last type of intermolecular force.

`0:43:53`

V3

0:45:00 **SECTION B: PHASE DIAGRAMS**

A **phase change** is the transformation of a substance from one state to another. When ice melts, water is changing from its solid phase to its liquid phase. As water boils, it goes from its liquid phase to vapor, its gaseous phase.

Solid → Liquid   *Liquid → Gas*

A PHASE DIAGRAM GIVES A NICE GRAPHICAL SUMMARY OF A SUBSTANCE'S PHASES.

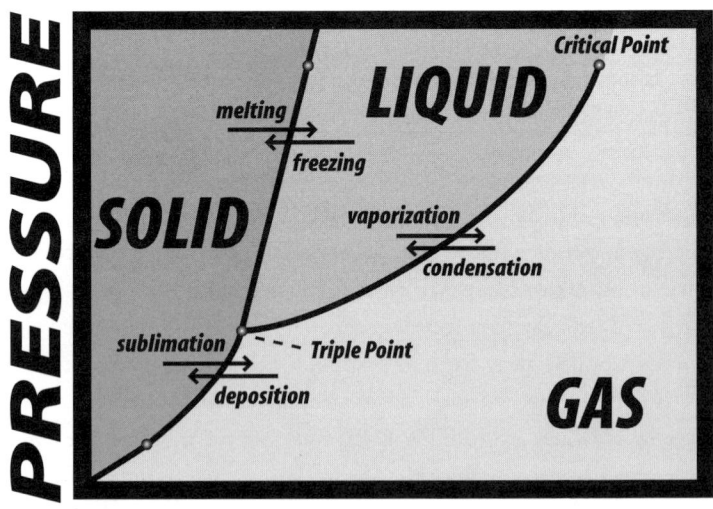

VIDEO NOTES

The Super-Charged World of Chemistry Part 3

A phase diagram shows which phase a substance is in at a particular temperature and pressure. At low temperature and high pressure, substances tend to be in their solid phase. At high temperature and high pressure, substances are generally liquid; and at high temperature and low pressure, substances are usually gaseous.

When two opposing processes occur at the same time––for example, if a liquid is evaporating and condensing simultaneously at the same rate––the substance is at *equilibrium*. The lines on the phase diagram represent the equilibrium point between states. The intersection of the three lines on the diagram is called the **triple point**. This point represents the pressure and temperature at which all three phases are at equilibrium. At the specific pressure and temperature indicated by the triple point, a substance moves between its liquid, gas, and solid phases simultaneously and at the same rate.

V3

SECTION C: VAPOR PRESSURE

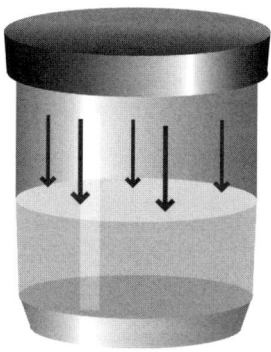

In a closed system, **vapor pressure** is the constant force that a vapor exerts above a liquid when equilibrium is established. So if you put some liquid in a jar and put a lid on it, at the appropriate temperature and pressure, the liquid will start turning to vapor inside the jar. After a while, the vapor will begin condensing again, and the system will reach equilibrium between the liquid and vapor phases—the substance will be condensing into liquid and evaporating into gas at the same time and at the same rate. Once equilibrium is established we can measure the vapor pressure.

The magnitude of the vapor pressure depends on how many molecules can shift into the gas phase—more gas molecules means higher vapor pressure. The number of molecules that can become gas depends on the temperature and strength of the intermolecular forces that hold the molecules together. For example, the hydrogen bonds that hold water molecules together are pretty strong, so water molecules don't escape into vapor phase very easily. The forces holding gasoline molecules together, on the other hand, are much weaker. You can tell that gasoline vaporizes easily, because you can smell gas vapor when you fill your tank.

Two out of three people inter-viewed admitted that they like the smell of gasoline when they're pumping gas into their car. The third person said that's why he hangs around gas stations.

FLASK OF FACTS:

+ Intermolecular forces are the forces of attraction between molecules: ion-dipole forces, dipole-dipole forces, London dispersion forces, and hydrogen bonds.

+ Intermolecular forces can cause gas molecules to stick together so that they become a liquid; intermolecular forces can also lock molecules rigidly in place so that they form solids.

+ Substances change phases at speci-fied temperatures and pressures, and when a substance is changing back and forth from one phase to another simultaneously and at the same rate, the substance is at equilibrium.

QUIZ 9

(ANSWERS ON PAGE 311)

1. Define the following terms.

    a. dipole-dipole forces

    b. hydrogen bonds

    c. London dispersion forces

    d. triple point

    e. phase change

    f. vapor pressure

2. Label points *a* through *g* on the following phase diagram, indicating carbon dioxide's phase at each of the points.

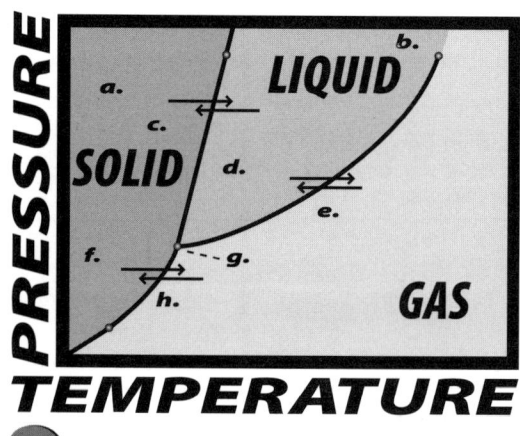

The Super-Charged World of Chemistry Part 3

PROPERTIES OF SOLUTIONS

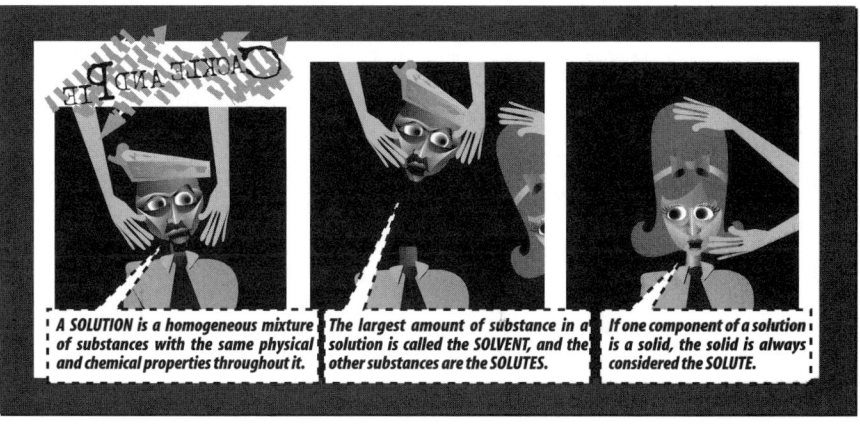

A SOLUTION is a homogeneous mixture of substances with the same physical and chemical properties throughout it.

The largest amount of substance in a solution is called the SOLVENT, and the other substances are the SOLUTES.

If one component of a solution is a solid, the solid is always considered the SOLUTE.

SECTION A: SOLUTION FORMATION

`0:49:45`

Liquids that can dissolve in each other are called **miscible**. However, not all liquids can combine to make a solution; those that are insoluble in each other are called **immiscible**. The general rule for solutions is *like dissolves like*. For instance, ionic and polar solutes are soluble in polar solvents, so the ionic solid NaCl (table salt) can dissolve in water (a polar liquid). Similarly, nonpolar solutes are soluble

`0:49:55`

"Mais non, of course you cannot dissolve oil in water, you are crazy? Eh? 'Ere is some magnifique nonpolar solvent for you: carbon tetra-chloride. MMM, délicieux."

only in nonpolar solvents. You know you can't dissolve cooking oil in water, right? That's because water is a polar solvent and cooking oil is a nonpolar substance. If you wanted to dissolve a nonpolar substance like oil, you'd have to get yourself some nonpolar solvent like carbon tetra-chloride.

`0:51:15` **Solubility** is the maximum amount of solute that can dissolve in a particular solvent at a particular temperature. Solubility varies depending on the temperature of the solution and the temperature of the solvent. If a solute is in its gas phase, then its solubility also depends on the pressure of the gas.

You've probably noticed you can only dissolve a certain amount of salt in a glass of water. If you keep adding salt to the water after you hit that saturation point, the salt won't dissolve; it will just lie in the bottom of the glass. However, if you change the temperature of the water or of the salt, the saturation point will change. If you raise the temperature

of the water, you can dissolve more salt in it. If you lower the temperature, you will usually reach the saturation point a lot sooner.

A **saturated solution** is a solution at equilibrium in which the maximum amount of solute has been dissolved; if you add any more solute to the solution, it will remain undissolved. When the proportion of solute to solvent is large, the solution is considered **concentrated**. A **dilute solution** has a small amount of solute compared with the amount of solvent.

`0:52:35`

`0:52:53`

When you create a solution by dissolving something in water, you've concocted an **aqueous solution**. Not all aqueous solutions are equal, however; some conduct electricity and some don't. It's true that water doesn't conduct electricity very well, but if your aqueous solution contains an ionic solid that will dissociate into ions, your solution will conduct electricity. The ionic solids that form these conductive solutions are called **electrolytes**.

Dissociation refers to a molecule **breaking up** into its **constituent** atoms. In this context, dissociation is when an **ionic solid** separates into **ions in solution**.

Naturally, solutions with a higher concentration of ions have greater electrical conductivity than those with a lower concentration of ions. We'll construct an experiment to demonstrate this, using the marvelous apparatus below.

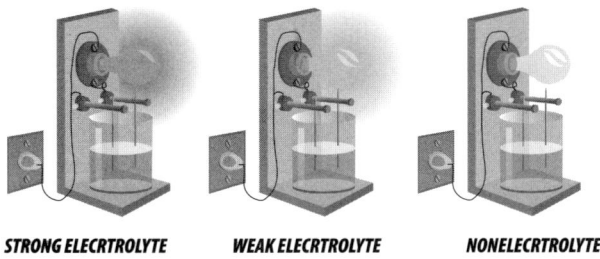

STRONG ELECRTROLYTE   **WEAK ELECRTROLYTE**   **NONELECRTROLYTE**

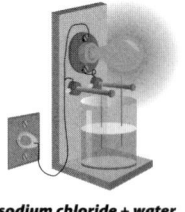

sodium chloride + water

If there are ions present in this solution, they will help conduct electricity between the electrodes stuck in the beaker. In short, the ions complete the electrical circuit. If we place a strong electrolyte like sodium chloride into the solution, it will ionize completely, the electricity will flow freely through it, and the light bulb will burn brightly.

vinegar + water

On the other hand, if we create a solution of acetic acid (vinegar) in water, the vinegar does not ionize completely, since it is only a weak electrolyte. Only some of the solute is ionic, so the solution is less conductive. Electricity does not move through the weak electrolyte as well, and consequently the light bulb does not burn as brightly.

The Super-Charged World of Chemistry Part 3

SECTION B: CONCENTRATION

`0:53:49`

When we're talking about the composition of a solution, we're not only referring to the components of the solution, but also the concentrations and ratios of those components. There are several ways of expressing the composition of a solution: mass percentage composition, mole fraction, molarity, and molality. Rest assured, we'll show you how to find them all––and even give you insight into the Art of Draining Your Spaghetti, plus Fourteen Ways to Cure the Blues––at no charge! (Well, outside the cost of this attractive and information-packed Study Sidekick, of course.)

The **mass percentage composition** of a solution measures the number of grams of solute in every 100 grams of solution. Two formulas will guide you through the realm of mass percentage:

`0:54:19`

$$\text{Mass percent of the solute} = \frac{\text{(mass of solute)}}{\text{(mass of solution)}} \times 100$$

mass of solution = mass of solvent + mass of solute

The first equation above shows that mass percentage equals the mass of the solute divided by the mass of the whole solution, times one hundred. The second equation tells you that the mass of the whole solution equals the mass of the solvent plus the mass of the solute.

What happens if you get a question like this in class?

Find the mass percentage composition of a solution containing 10 grams of NaCl (table salt) in 90.0 grams of water.

No sweat. Start by finding the mass of the solution, which should be a snap. 10 grams of salt plus 90 grams of water equals 100 grams. Right on. Now plug into the percentage composition formula: the percentage composition of salt in the solution equals 10 grams divided by 100 grams, times one hundred.

$$\text{Mass of the solution} = 90 \text{ grams } H_2O + 10 \text{ grams NaCl}$$
$$= 100 \text{ grams of solution}$$

$$NaCl\% = \frac{(10 \text{ grams})}{(100 \text{ grams})} \times 100 = 10.0\%$$

Now you can sharpen your pencil and write, "The mass percentage composition of the solution is 10%."

Doesn't that go down nice & smooth?

Refreshing, like an icy, frothy mug of lemonade.

The **mole fraction** of a compound is kind of like percent composition, because when you're looking at a certain part of a solution, the mole fraction will tell you what percent of the solution your component represents. The mole fraction of a compound, also known as X, is equal to the number of moles of the compound you're measuring, divided by the total number of moles of all components of the system. When you add up all the mole fractions in a system, your answer will always equal one. (If they don't, either you added wrong, or the nature of reality has shifted and the standard laws of physics no longer apply. Your life is now a Dalí painting!)

`0:55:19`

204

Let's take a second look at the example we just did to explain mass percent composition; but this time we'll find the mole fraction of NaCl in the solution. First we must convert from grams to moles, since mole fractions are measured in moles. (You can refer back to the simplified version of the conversion equation we talked about earlier in section E of *The Super-Charged World of Chemistry, Part 1.*)

We divide the mass of the sample, 10.0 grams, by the formula weight of NaCl. Remember to use the formula weight for NaCl because it is a crystal, not a molecule. The formula weight of one mole of NaCl is the sum of the masses of the atoms in NaCl's empirical formula: the mass of sodium (23.0 g/mol) plus the mass of chlorine (35.5 g/mol) equals 58.5 g/mol. Divide 10.0 grams by 58.5 grams and you get 0.171 moles of NaCl.

$$n_{NaCl} = \frac{10 \text{ grams}}{(23.0 + 35.5)}$$

$$= 0.171 \text{ moles of NaCl}$$

We'll follow the same principle to work out how many moles of water we have. Since water is a molecule, we use molecular weight this time: that's two times the atomic mass of hydrogen plus the atomic mass of oxygen, which is 18.02 grams per mole. 90.0 grams divided by 18.02 grams equals 5.00 moles of water.

$$\text{mole fraction of a component} = \frac{\text{moles of component}}{\text{total moles of all components}}$$

$$nH_2O = \frac{90.0 \text{ g}}{[2(1.008)+16.00] \text{ g}}$$

$$nH_2O = \frac{90.0 \text{ g}}{18.02 \text{ g}} = 5.00 \text{ moles}$$

Now we can put all these pieces together and find the mole fraction (X) of NaCl. According to our mole fraction formula, we divide the number of moles of NaCl by the total number of moles of solution we have. Our solution is 0.171 moles plus 5.00 moles, which yields a total of 5.17 moles. We divide 0.171 by 5.17 moles, and get 0.0331 for the mole fraction for HCl.

$$\text{mole fraction of a component} = \frac{\text{moles of component}}{\text{total moles of all components}}$$

$$\text{mole fraction (X)} = X_{NaCl} = \frac{0.171 \text{ moles}}{(0.171+ 5.00) \text{ moles}}$$

$$= \frac{0.171 \text{ moles}}{5.17 \text{ moles}}$$

$$= 0.0331$$

Chemster

Doing the same calculation to find the water's mole fraction, we divide 5.00 moles by 5.17 moles and get 0.967. So the mole fraction (X) for water is 0.967. And here you'll notice the wonky thing about mole fractions--they're always expressed as a decimal, despite the name mole *fraction*. And when in Chemland, we do as the Chemsters do. So we'll stick to decimals when we're talking about mole fractions.

$$X_{\frac{1}{2}-\frac{2}{3}} = \frac{5.00 \text{ moles}}{5.17 \text{ moles}}$$

$$X_{\frac{1}{2}-\frac{2}{3}} = 0.967$$

`0:58:17`

Time to hike up your trunks and cannonball into molarity. The **molarity** of a solution, represented by a capital M, is the number of moles of a solute in one liter of a solution. This is another way of looking at concentration in a solution, but this time it's not in terms of percentages or fractions of the whole solution. Molarity simply measures the amount (in moles) of a certain solute in one liter of the solution.

$$M = \frac{n, \text{ moles of solute}}{v, \text{ liters of solution}}$$

The Super-Charged World of Chemistry Part 3

Want to do an example? Y/N

• If you answered *yes*: Continue reading the next paragraph, Chemsters.

• If you answered *no*: Flip directly to the Stress Relief section.

• If you answered *maybe*: Quit teasing, you tease.

Here's the example:

Find the molarity of 20.0 grams of sodium chloride in a 100 milliliter solution.

As always, we must first convert from grams to moles; so we divide the mass of our sample by the formula weight of our NaCl.

$$n = \frac{20\ g}{58 g/mol} = 0.34\ \text{moles}$$

20 grams of NaCl divided by the formula weight of NaCl (which we determined earlier: 58 g/mol) gives us 0.34 moles of NaCl. Now it's time to whip out our spankin' new molarity formula. Molarity equals the number of moles of solute divided by the volume in liters of solution. Since our quantities are in mil-

V3

liliters, we've got to translate into liters: 100 milliliters equals 0.100 liters. We set up our equation so that 0.34 moles is divided by 0.100 liters of solution. The molarity of our solution is 3.4.

$$M = \frac{n, \text{moles of solute}}{V, \text{liters of solution}}$$

$$100 \text{ mls} = 0.100 \text{ L}$$

$$M = \frac{0.34 \text{ moles}}{0.100 \text{ Liters}} = 3.4 \text{ M}$$

ONTO MOLALITY!
But heads up on this one!
Make sure you don't get molarity and molality confused!

The **molality** of a solution, represented by a small m, is the number of moles of solute per kilogram of solvent. It's the amount of solute in one little kilogram of solvent alone, not in the whole solution. For our NaCl example, the molality is the amount of salt in *one kilogram of the water*, while molarity is the number of moles of solute *per liter of the whole solution*. Got it?

1:00:00

The equation for molality is the following:

$$m = \frac{n, \text{moles of solute}}{\text{kilograms of solvent}}$$

The Super-Charged World of Chemistry Part 3

Let's go back to our golden example and find the molality of the solution (10.0 grams of NaCl in 90.0 grams of water). You know what to do first: find the number of moles of solute you have. Set up your equation so that n equals 10 grams of NaCl divided by the formula weight of NaCl. We'll take the formula weight of NaCl to one decimal place further in this example than we did when we were finding molarity, because this time we're using three significant figures in our calculations. As before, we get 0.171 moles of NaCl.

$$m = \frac{n, \text{ moles of solute}}{\text{kilograms of solvent}}$$

$$n = \frac{10.0g}{58.5 \text{ g/mol}}$$

$$= 0.171 \text{ g/mol of NaCl}$$

Go, and never darken my towels again.

– Groucho Marx

Now we're ready to plug the values we've found into the molality equation. Molality equals the number of moles of solute divided by kilograms of solvent. We have 90 grams of water (the solvent) and 90 grams is the same thing as 0.0900 kilograms. So the molality equals 0.171 moles divided by 0.0900 kilograms, which works out to 1.90.

$$\text{kg of solvent} = \frac{90.0g}{1000 \text{ g}/1 \text{ kg}} = 0.0900 \text{ kg}$$

$$m = \frac{\text{moles of solute}}{\text{kg of solvent}}$$

moles of solute = 0.171 moles

kilograms of solvent = 0.0900 kilograms

$$m = \frac{0.171 \text{ mol}}{0.0900 \text{ kg}} = 1.90 \text{ m}$$

VIDEO NOTES

The Super-Charged World of Chemistry Part 3

Mole of Knowledge:

- Liquids that can dissolve in each other are miscible, and liquids that can't dissolve in each other are immiscible.

- Solubility refers to the maximum amount of a solute that you can dissolve in a solvent.

- A saturated solution is a solution that has accepted as much of a solute as it can at a particular temperature.

- Mass percentage composition, mole fraction, molarity, and molality are all different ways to describe the composition of solutions.

If Today Was a Fish, I'd Throw It
Back In.

– Song title

`1:03:22` **SECTION C: COLLIGATIVE PROPERTIES**

Colligative properties are properties determined by the concentration (quantity) of solute in a solution. These properties, which include freezing-point depression, boiling-point elevation, and osmosis, do not have to do with the *kind* of solute we're dealing with––just *how much* of it is present.

Freezing-point depression and boiling-point elevation both alter the temperature at which a change of state occurs, so we'll take a look at these two properties together.

When solute particles interfere with a solution's ability to freeze, it delays the crystal formation that occurs during freezing and lowers (or depresses) the freezing `1:04:10` point of the solution. The **freezing-point depression,** represented by Δt_f, is proportional to the molality of the solute. To calculate freezing-point depression, multiply the molality of the solution by the molal freezing-point-depression constant, K_f. You can look up specific K_f values in your textbook.

$$\Delta t_f = K_f m$$

The antifreeze in your car is a solution of ethylene glycol in water. The solute, ethylene glycol, interferes with the ability of the water to form ice crystals. This has a practical benefit––it freezes at a lower temperature than plain old water, so it's great for the radiator of your car.

ANTIFREEZE IS TOPS. DON'T DRIVE THAT DUNE-BUGGY WITHOUT IT.

Antifreeze also has the virtue of a higher boiling point than water. Because of this, antifreeze makes a great substitution for water in your radiator––it allows the engine to operate at a higher temperature than it would otherwise. Since a liquid boils when its vapor pressure equals the atmospheric pressure, the addition of some solvents will cause a solution to boil at a higher temperature. When a solute raises the boiling point of a solution, this is called **boiling-point elevation**. The formula for calculating boiling-point elevation is $\Delta t_b = K_b m$. Look familiar? Δt_b represents the boiling-point elevation, K_b is the molal boiling-point-elevation constant, and m is the molality of the solution.

1:05:52

Let's tackle an example to illustrate both freezing-point depression and boiling-point elevation.

Automobile antifreeze is made up of ethylene glycol, $C_2H_6O_2$, and water. Calculate both the freezing-point depression and boiling-point elevation of a 24.1 mass percent solution of antifreeze in water.

Since you remember that mass percent is the number of grams of solute in each 100 grams of solution, you know there are 24.1 grams of ethylene glycol in 75.9 grams of water. Great––you've got the number of grams of solute and solvent just from reading the problem.

We'll start with the freezing point, using the equation $\Delta t_f = K_f m$. The first thing to do is to figure out the components of this equation. Remember that m equals the number of moles of solute divided by the kilograms of solvent. To find the molality, we convert our 24.1 grams of solute into moles, using the quick formula we discussed earlier. That's 24.1 grams divided by the weight of 1 mole of ethylene glycol (62 g/mol), which gives us 0.389 moles of ethylene glycol.

$$n = \frac{24.1 \text{ g}}{62.0 \text{ g/mol}} = 0.389 \text{ moles}$$

Now we can find the molality: we divide 0.389 by 75.9 grams (or 0.0759 kilograms) to get 5.12 m.

$$m = \frac{\text{moles of solute}}{\text{kg of solvent}}$$

$$m = \frac{0.389 \text{ mol}}{0.0759 \text{ kg}} = 5.13 \text{ m}$$

Dipping into the well of information that is our chemistry textbook, we find that K_f, the molal freezing-point-depression constant for water is 1.86. We multiply that value by the molality, 5.09, and get a freezing-point depression of 9.5 degrees.

$$K_f (H_2O) = 1.86$$

$$\Delta t_f = 1.86 \times 5.09 \text{ m} = 9.5$$
$$\Delta t_f = 9.5$$

To finish off the calculation, we subtract the freezing-point depression value, 9.5, from the usual freezing point of the solvent, so that we get the freezing point of our new solution of water plus antifreeze. Since water freezes at 0.0°C, we subtract 9.5°C. The freezing point of our solution is −9.5°C.

$$0.0°C - 9.5°C = {}^-9.5°C$$

So the answer to the first part of the question is that to freeze a 24.1 mass percent solution of water and antifreeze, you must lower (depress) the temperature of the solution by 9.5°C.

V 3

Fourteen Ways to Cure the Blues

1. Sleep. Eat. Repeat as necessary.

2. Hit the road. Tell your parents you're going to discover America and your beatnik soul.

3. Dye your hair.

4. Pierce something.

5. Re-upholster your personality.

6. Listen to zydeco.

7. Hunt down a rat in your dorm, teach it to swallow fire, and go on tour.

8. Watch an MTV beach party marathon.

9. Walk into a class that's taking a test. Act like you're in the class and take the test, too. After five minutes, get up with a stretch and a yawn, saying, "Boy, was that a breeze!" Then walk out.

10. Read your roommate's diary. Make corrections.

11. Order a pizza. Order one hundred pizzas. (Make sure you order them on someone else's meal card.)

12. Streak.

13. Buy a few dozen Study Sidekicks. After memorizing them all, build a fort out of them.

14. Press the eject button.

VIDEO NOTES

The Super-Charged World of Chemistry Part 3

We can find the new boiling point of the solution in pretty much the same way. We use the same molality, 5.09, and multiply it by the molal boiling-point-elevation constant, which our table says is 0.52 for water. This gives us a boiling-point elevation of 2.6 degrees. We add 2.6 to the boiling point of water, 100°C, and get an elevated boiling point of 102.6°C.

$$\Delta t_b = K_b m$$

$$\Delta t_b = 0.552 \times 5.09$$

$$\Delta t_b = 2.6\ °C$$

$$100.0\ °C + 2.6\ °C = 102.6°C$$

SECTION D: OSMOSIS

Osmosis is the process in which solvent flows through a semipermeable membrane from a dilute solution to a more concentrated one. Osmosis is considered a colligative property because it depends on the number of solute molecules or ions in the solution.

V 3

The Art of Draining Your Spaghetti

A semipermeable membrane is a divider of sorts, similar to a sieve with very, very tiny openings that only allow some things to pass through. If you've ever drained spaghetti with a colander, you've counted on the colander to allow the water through while holding onto the pasta. This is a large-scale version of osmosis: some stuff is too big to get through the pores or little holes and some stuff passes through easily.

When you put a semipermeable membrane between two solutions of different concentrations and the solvent is able to pass through the membrane, osmosis will occur. The solvent in the solution of lower concentration will pass through the membrane to dilute the solution of higher concentration until the two solutions are more equal in concentration.

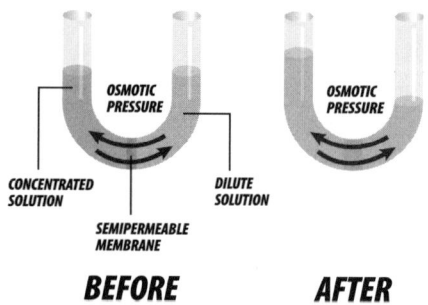

OSMOTIC PRESSURE

OSMOTIC PRESSURE

CONCENTRATED SOLUTION

DILUTE SOLUTION

SEMIPERMEABLE MEMBRANE

BEFORE   **AFTER**

Look at it this way: if the solvent is bopping back and forth through the membrane, but the solute is too big to pass through the membrane, it makes sense that the solution with more solvent and less solute will get through more quickly than the solution that has more solute in it.

VIDEO NOTES

The Super-Charged World of Chemistry Part 3

Let's make this a wee bit clearer. Say we have a dilute solution and a concentrated solution, both made of table salt and water. The two solutions are in a U-shaped tube and are separated by a semipermeable membrane. The molecules of solute (salt) are too large to pass through the membrane, but the water molecules can pass through easily. The more concentrated solution has more salt floating around in it, and the salt gets in the way of the water molecules' passage through the membrane. The dilute solution has less salt in the way, so it passes through the membrane more easily.

The water in the dilute solution will make its way through the membrane more quickly than the water in the concentrated solution, so the concentrated solution is going to have more solvent flowing *into it* than the dilute solution will. Since the solutions are striving toward equilibrium, the solution on the left needs more solvent to have the same concentration as the solution on the right. Therefore the volume of the solution on the left will increase to accommodate some of the solvent that was originally on the right.

Keep in mind that the solvent moves toward the more highly concentrated solution.

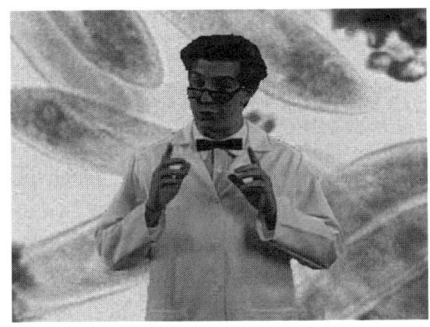

Anyone know what this is?

Class?

Anyone?

Anyone?

– *Ferris Bueller's Day Off*

V 3

DEIONIZED INFORMATION:

- Freezing-point depression, boiling-point elevation, and osmosis are all considered colligative properties, since they depend only on the concentration of a solution, and not on what the solute is composed of.

- Freezing-point depression calculates how much a solute interferes with a solvent's ability to freeze––that is, how much the temperature of the solvent must decrease before it reaches its freezing point.

- Boiling-point elevation calculates how much a solute interferes with a solvent's ability to boil––that is, how much the temperature of the solvent must increase before it will boil.

- Osmosis is the movement of solvent through a semipermeable membrane from a solution of lower concentration to a solution of higher concentration.

QUIZ 10

(ANSWERS ON PAGE 313)

1. Define the following terms.

   a. miscible

   b. saturated solution

   c. molality

   d. mole fraction

   d. molarity

   e. osmosis

2. If a solution is made up of 31.2 grams of potassium nitrate (KNO_3) and 100.0 grams of water, calculate the mole fraction of KNO_3 and the molality of the solution.

3. If the density of the solution is 1.00 g/ml, what is the molarity of the solution?

4. If the molality of an aqueous sugar solution is 0.250 m, what is the boiling point and freezing point of this solution? (For water, $K_f = 1.86$ and $K_b = 0.521$.)

5. How would you prepare 130 grams of an aqueous solution whose mass is 3.05% $MgCl_2$?

QUIZ XYZ

1. How do you pronounce "cation"?

    a. cay-shun

    b. cat-eye-on

    c. katey-on

2. What is a "hippocampus"?

    a. an institution of higher learning for thick-skinned, four-toed, aquatic mammals

    c. a venue for chariot races in ancient Greece

    d. part of your brain

3. Which of the following are actual names of towns in West Virginia?

    a. Cornstalk (yes / no)

    b. Crumb (yes / no)

    c. Catastrophe (yes / no)

    d. Accident (yes / no)

Tickle your noggin with Practice Exam 4.

OTHER IMPORTANT STUFF

STUFF Nº 1: MOLECULAR FORMULAS

In *The Super-Charged World of Chemistry, Part 1*, we learned how to determine empirical formulas. We can also find the molecular formula of a compound if we know both the empirical formula and the molar mass of the compound.

What's the difference between an empirical formula and a molecular formula? The bottom line is that the empirical formula indicates the ratio of the atoms in the compound, whereas the molecular formula indicates the exact number of each atom in the compound. The molecular formula is a multiple of the empirical formula.

> True love is the best thing in the world, except for a nice MLT, mutton, lettuce, and tomato sandwich...where the mutton is nice and lean, and the tomato's ripe.
>
> – Miracle Max in *The Princess Bride*

Suppose we come across a compound with a percent composition of 30.4% nitrogen and 69.6% oxygen, and a molar mass of 92 g/mol. What is the molecular formula of the compound?

The first thing to do is roll up your sleeves and find the empirical formula of the compound: the ratio of moles of nitrogen to moles of oxygen. (This calculation is just like the one we did in *Part 1* to find the empirical formula of water.)

AW = atomic weight in grams per mole

$$\text{moles of } \textbf{\textit{nitrogen}} = \frac{1 \text{ mol N}}{\text{AW of N}} \times \text{weight of sample of N (g)}$$

$$\text{moles of } \textbf{\textit{nitrogen}} = \frac{1 \text{ mol N}}{14.0 \text{ g/mol}} \times 30.4 \text{ g N} = 2.17 \text{ mol}$$

$$\text{moles of } \textbf{\textit{oxygen}} = \frac{1 \text{ mol of O}}{\text{AW of O}} \times \text{weight of sample of O (g)}$$

$$\text{moles of } \textbf{\textit{oxygen}} = \frac{1 \text{ mol O}}{16.0 \text{g/mol}} = 69.6 \text{ g} = 4.35 \text{ mol}$$

The ratio of oxygen to nitrogen is $^{4.35}/_{2.17}$, which reduces to 2:1. This ratio tells us that we have two atoms of oxygen for every atom of nitrogen. Therefore, we can write the empirical formula of the compound like this: NO_2.

Our compound's molecular formula will be a multiple of its empirical formula, reflecting the actual number of atoms present in the compound: $(NO_2)x$. We need to determine x, which we'll do using the following formula:

$$x = \frac{\text{molar mass}}{\text{empirical weight}}$$

$$\text{molar mass} = 92.0 \text{ g/mol}$$

$$\text{empirical weight} = (\text{AW of N}) + (2 \times \text{AW of O})$$
$$= 14.0 + (2 \times 16.0)$$
$$= 46.0 \text{ g/mol}$$

$$x = \frac{92.0 \text{ g/mol}}{46.0 \text{ g/mol}} = 2$$

We multiply our empirical formula by our x value, 2, to get the molecular formula of the compound: $(NO_2)_2$ or N_2O_4. That's all there is to it.

226

Formula for Freshman-15 Weight:

X = # of slices of pizza eaten past 3:00 a.m. +
# of cans of Mountain Dew drunk daily +
# of chocolate-coated snacks per week +
7 (college growth constant)

Y = (dining-hall frozen yogurt per day +
volume of diet sodas) × 18 (saccharin constant)

$$\frac{\text{high school weight} + X}{\text{college weight} - Y} = \text{Freshman 15 (average weight gained during first year of college)}$$

Boy, the things I do for England.

– Prince Charles, on sampling
snake meat

STUFF N⁰. 2: NOMENCLATURE: A QUICK GUIDE TO THE NAMING OF SIMPLE COMPOUNDS

A. Naming Binary Ionic Compounds

Binary ionic compounds are composed of two elements: a cation (a positively charged ion) and an anion (a negatively charged ion). So how do you know what to call these compounds? The following three rules of thumb will help you call binary compounds by their proper names.

1. The cation is always named first, then the anion.

2. The cation's name is that of the element. For example, Ca^{++} is just called calcium.

3. To get the anion's name, take the first part of the element's name and add *-ide*. For example, S_2- is called "sulfide."

IS

Here are some examples:

COMPOUND	NAME
MgO	magnesium oxide
NaCl	sodium chloride
KI	potassium iodide
CaS	calcium sulfide

And voilà you're speaking Chemistry.

"What kind of plane is it?"

"Oh, it's a big, pretty, white plane with red stripes, curtains in the windows, and wheels, and it looks like a big Tylenol."

– *Airplane!*

B. More Binary Compounds

Some binary compounds are composed of cations that can form more than one type of ion. For example, copper can form Cu I and Cu II. When you're naming these cations, make sure you indicate the charge with a Roman numeral. In the case of the copper cations, Cu I is written as Copper (I), and Cu II is Copper (II). (Note: If an element forms only one type of cation, you don't have to bother with a Roman numeral.)

Here are some examples:

COMPOUND	NAME
$FeCl_3$	iron (III) chloride
$AlCl_3$	aluminum chloride
(aluminum only forms one cation)	
$MnBr_2$	manganese (II) bromide

IS

C. Polyatomic Ions

The table below lists the common polyatomic ions and their names.

NAME	FORMULA
Mercury(I) (or mercurous)	Hg_2^+
Ammonium	NH_4^+
Cyanide	CN^-
Hydrogen Carbonate (or bicarbonate)	HCO_3
Acetate	$C_2H_3O_2^-$
Oxalate	$C_2O_4^{2-}$
Hypochlorite	ClO^-
Chlorite	ClO_2^-

NAME	FORMULA
Chlorate	ClO_3
Perchlorate	ClO_4
Chromate	CrO_4^{2-}
Dichromate	$Cr_2O_7^{2-}$
Permanganate	MnO_4^-
Nitrite	NO_2^-
Nitrate	NO_3^-
Hydroxide	OH^-
Peroxide	O_2^-
Phosphate	PO_4^{3-}
Monohydrogen phosphate	HPO_4^{2-}
Dihydrogen phosphate	$H_2PO_4^-$
Sulfite	SO_3^{2-}

NAME	FORMULA
Sulfate	SO_4^{2-}
Hydrogen sulfite (or bisulfite)	HSO_3^-
Hydrogen sulfate (or bisulfate)	HSO_4^-
Thiosulfate	$S2O_3^{2-}$

Here are some real-life examples of compounds containing the polyatomic atoms listed in Table I:

COMPOUND	NAME
$Na_2Cr_2O_7$	sodium dichromate
$NH_4C_2H_3O_2$	ammonium acetate
K_2SO_3	potassium sulfite

IMPORTANT STUFF

The Super-Charged World of Chemistry Parts 1, 2 & 3

D. Binary Covalent Compounds

Binary covalent compounds are compounds that form between nonmetals. There are three simple rules for naming these types of compounds:

NONMETALS ARE THOSE ELEMENTS TO THE RIGHT OF THE DIAGONAL LINE ON THE PERIODIC TABLE.

1. Name the element listed first in the compound, using its element name.

2. Name the second element as if it were an anion.

3. Use appropriate prefixes to indicate the number of atoms (see table below).

PREFIX	NUMBER OF ATOMS
mono-	1
di-	2
tri-	3
tetra-	4
penta-	5
hexa-	6
hepta-	7

IS

Let's put this into practice. Here are some of the oxides of nitrogen.

BINARY COVALENT COMPOUND	NAME
NO	nitrogen monoxide *
N_2O	dinitrogen oxide
NO_2	nitrogen dioxide
N_2O_5	dinitrogen pentoxide

(* never use *mono-* with the first element)

E. Acids

When you dissolve certain compounds in water, they become acids. Happily, there is a systematic method for naming this kind of compound. Isn't that nice?

• If the acids do not contain oxygen, they are named in the following manner:

hydro + {name of the anion ending in *-ic*} acid.

For example, HCl (aq) is "hydro" + "chloric" + "acid," which becomes "hydrochloric acid."

COMPOUND	NAME
HI (aq)	hydroiodic acid
H_2S (aq)	hydrosulfuric acid
HCN (aq)	hydrocyanic acid *

(* CN^- is a polyatomic ion from Table I)

- When the acid's anion does contain oxygen, these rules apply:
- If the name of the anion ends in *-ate*, then the acid name ends in *-ic*.
- If the anion ends in *-ite*, then the acid name ends in *-ous*.

Here are some examples to tickle and delight you.

COMPOUND	NAME
H_2SO_3	sulfuric acid
H_2CO_3	carbonic acid
HNO_2	nitrous acid
$HClO_2$	chlorous acid

F. *Extraneous caffeinated nomenclature:*

COFFEE	NAME
espresso	Françoise
cappuccino	Paolo
macchiato	María Conlatte
American	Joe

"What is the air speed of an unladen swallow?"

"An African or European swallow?"

"Yaaaaaaah!"

– *Monty Python and the Holy Grail*

IMPORTANT STUFF

STUFF N⁰. 3: EXCEPTIONS TO THE OCTET RULE

In *The Super-Charged World of Chemistry Part 2*, we mentioned that there are a few exceptions to the octet rule when you're drawing Lewis structures. Specifically, compounds consisting of elements from groups II and III in the periodic table form electron-deficient compounds (compounds with less than eight electrons).

The compounds BeH_2 and BF_3 demonstrate this phenomenon. Take a look at their Lewis structures:

Beryllium has only 4 electrons around it rather than the usual 8, and boron has only 6 rather than the usual 8. Since these two elements have less than 8 electrons, we call them electron-deficient compounds.

You may come across some atoms with more than 8 electrons surrounding them when you're drawing Lewis structures. PF_5 and SiF_6^{2-} are examples of this type of compound:

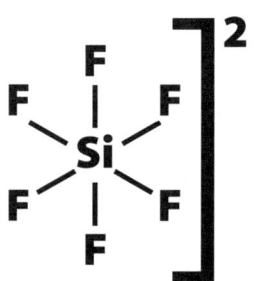

Since the P atom in PF_5 has 10 electrons surrounding it and the Si atom in the SiF_6^{2-} ion has 12 electrons surrounding it, these atoms have an expanded valence. A good way to check whether or not an atom has an an expanded valence is to check its n (principle quantum number), because an atom can only have an expanded valence when its n value is greater than or equal to 3.

We've got armadillos in our trousers.
It's quite frightening.

– *This Is Spinal Tap*

STUFF 4: FORMAL CHARGE

Formal charge is a concept that helps us out when two or more Lewis structure forms exist and we're trying to choose the most likely one. It is a method of assigning charges to a molecule or ion, given that the constituent atoms share the bonding electrons equally. We can determine a molecule's formal charge by assigning electrons according to the following guidelines:

1. Half of the electrons in a bond are assigned to each atom in the bond.

2. Both electrons in a lone pair are assigned to the atom to which the lone pair belongs.

After assigning electrons, we can calculate the formal charge using this handy-dandy formula:

Formal charge = valence electrons of the free atom $-\frac{1}{2}$ (# of electrons in a bond) $-$ (# of lone-pair electrons)

$$\overset{\ominus}{:\overset{..}{O}} - \overset{\oplus}{N} = \overset{..}{\overset{..}{O}}:$$
$$|$$
$$:\overset{..}{O}:$$
$$\ominus$$

Let's get to the heart of this formal charge business. Consider the compound NO_3-. How do we go about this? Draw the Lewis structure, then assign formal charges. Right-o.

If you can write several Lewis structures for a molecule, the more preferable structure is the one with the smallest formal charge (or the formal charge on the most electronegative atom).

$$\ominus:\overset{..}{\overset{..}{O}} \leftarrow \overset{\text{-1}}{S} = O$$
$$\|$$
$$O$$

Let's take SO_3 as an example. We'll draw the Lewis structures, calculate the formal charge, and then determine the most likely structure.

$$\overset{\text{-1}}{\ominus}:\overset{..}{\overset{..}{O}} \leftarrow \overset{\text{-2}}{S} = O$$
$$\downarrow$$
$$:\overset{..}{O}:$$
$$\overset{\text{-1}}{}$$

A and B have formal charges on the atoms whereas structure C does not. Therefore, C is our Lewis structure of choice.

$$:\overset{..}{\overset{..}{O}} = S = \overset{..}{\overset{..}{O}}:$$
$$\|$$
$$:\overset{..}{O}:$$

STUFF N⁰. 5: MOLECULAR GEOMETRY

In *The Super-Charged World of Chemistry Part 3*, we looked at how the VSEPR model predicted the geometry of molecules up to AX₄. Now we'll take you a step further and consider categories of five and six electron pairs (AX₅ and AX₆). **Woohoo!**

The Lewis structure for the compound PF₅ shows us that it has five bonding pairs that form a trigonal bipyramid.

In any molecule that has some sort of combination of 5 lone pairs and bonding pairs, the lone pairs will always reside in the trigonal plane of the trigonal bipyramid.

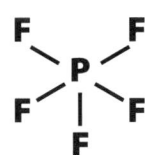

Check out the Lewis structure for the compound SiF₆²⁻. The diagram shows us that there are 6 bonding pairs of electrons and that these 6 pairs are arranged around the silicon atom in an octahedral fashion.

IS

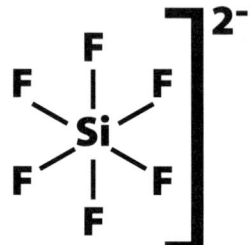

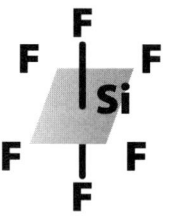

In molecules with some combination of 6 lone pairs and bonding pairs of electrons, the lone pairs of electrons take up positions as far away from each other as possible.

When we discussed orbital overlap in *Part 3*, we looked at the hybrid orbitals up to sp^3, or the hybrid orbital numbers 1 through 4. Just to flesh out our understanding of hybrid orbitals, we'll do one more example: the hybrid orbital number 6. This hybrid orbital number gives rise to $d^2 sp^3$ hybridization and an octahedral arrangement around the central atom. SF_6 is an example of this type of hybridization.

We also talked about molecular orbitals for hydrogen and helium in *Part 3*. Now we'll examine the molecular orbitals for the second-row, diatomic atoms. In these molecules, molecular orbitals are formed by the overlap of atomic orbitals with values of $n = 1$ and $n = 2$.

When $n = 1$, the 1s atomic orbitals overlap to form a σ_{1s} bonding molecular orbital and a σ_{1s} antibonding molecular orbital. These orbitals result from the direct, head-on overlap of atomic orbitals and are called sigma bonds. When $n = 2$, we also get head-on overlap of atomic orbitals between the $2s$ and the $2p_x$ orbitals, forming σ_{1s} and σ_{2p} bonding and antibonding molecular orbitals.

When the $2p$ orbitals perpendicular to the x-axis overlap (the py and the pz orbitals), they do so in a side-to-side manner. Just as with the s orbitals, both bonding and antibonding orbitals are produced. The order of energy of these orbitals is $\sigma_{2s} > \sigma^*_{2s} > \pi_{2py}, \pi_{2pz} > \sigma_{2px} > \pi^*_{2py}, \pi^*_{2pz} > \sigma^*_{2px}$. Here's a molecular orbital diagram that shows how the orbitals relate to each other in terms of energy.

ATOMIC ORBITALS	*MOLECULAR ORBITALS*	ATOMIC ORBITALS

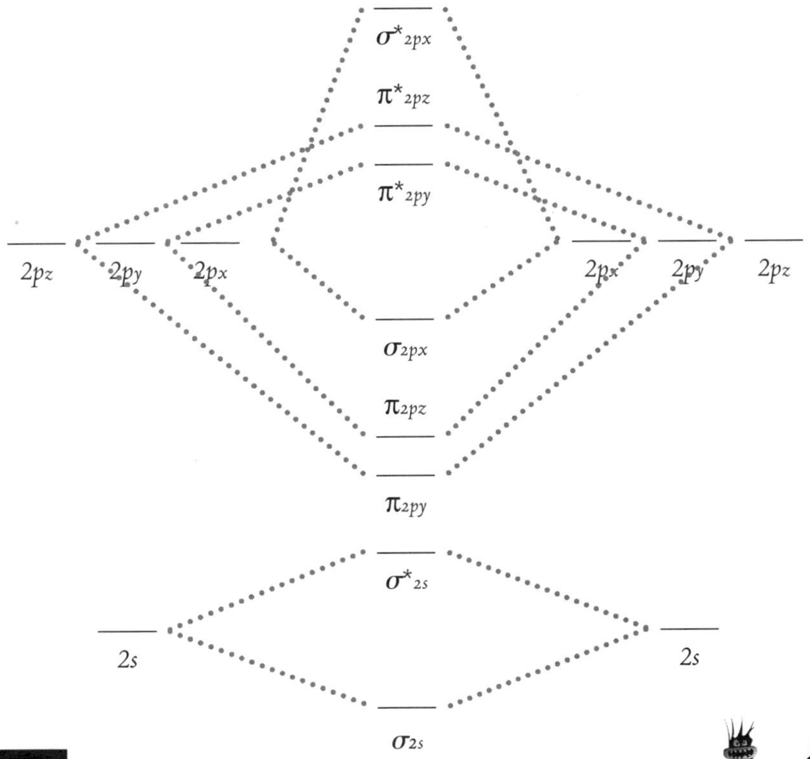

$$\sigma^*_{2px}$$

$$\pi^*_{2pz}$$

$$\pi^*_{2py}$$

$2pz$ $2py$ $2px$ $2px$ $2py$ $2pz$

$$\sigma_{2px}$$

$$\pi_{2pz}$$

$$\pi_{2py}$$

$$\sigma^*_{2s}$$

$2s$ $2s$

$$\sigma_{2s}$$

1S

244

Like all other orbitals, each molecular orbital can only accommodate two electrons. You can predict the electronic configuration for the second-row, diatomic molecules just as you did for atoms and ions in Part 2. You can also calculate their bond order by using the relationship we presented earlier:

$$\text{Bond order} = \frac{\text{\# of bonding electrons} - \text{\# of antibonding electrons}}{2}$$

The table below shows the electron configurations, bond order, bond length, bond energy, and magnetic properties of several molecules.

Molecule	Electron configuration	Bond Order
Li_2	$(\sigma_{2s})^2$	1
Be_2	$(\sigma_{2s})^2(\sigma^*_{2s})^2$	0
B_2	$(\sigma2s)^2(\sigma^*_{2s})^2(\pi_{2py})^1(\pi_{2pz})^1$	1
C_2	$(\sigma_{2s})^2(\sigma^*_{2s})^2(\pi_{2py})^2(\pi_{2pz})^2$	2
N_2	$(\sigma_{2s})^2(\sigma^*_{2s})^2(\pi_{2py})^2(\pi_{2pz})^2(\sigma_{2px})^2$	3
O_2	$(\sigma_{2s})^2(\sigma^*_{2s})^2(\pi_{2py})^2(\pi_{2pz})^2, (\sigma_{2px})^2(\pi^*_{2py})^1(\pi^*_{2pz})^1$	2
F_2	$(\sigma_{2s})^2(\sigma^*_{2s})^2(\pi_{2py})^2(\pi_{2pz})^2, (\sigma_{2px})^2(\pi^*_{2py})^2(\pi^*_{2pz})^2$	1
Ne_2	$(\sigma_{2s})^2(\sigma^*_{2s})^2(\pi_{2py})^2(\pi_{2pz})^2,(\sigma_{2px})^2(\pi^*_{2py})^2(\pi^*_{2pz})^2(\sigma^*_{2px})^2$	0

Looking closely at the table below, you'll notice several things:

- If a substance is paramagnetic, it has unpaired electrons and is attracted to a magnetic field.

- If a substance is diamagnetic, it has no unpaired electrons and is repelled by a magnetic field.

- As bond order increases, bond length decreases and bond energy increases.

Bond Length	Bond Energy	Magnetic Property
2.67	110	diamagnetic
	no stable molecule forms	
1.59	290	paramagnetic
1.24	602	diamagnetic
1.10	942	diamagnetic
1.21	494	paramagnetic
1.42	155	diamagnetic
	no stable molecule forms	

(The bond lengths are given in angstroms and the bond energies are given in kJ/mole.)

STUFF NO. 6: GASES

Let's talk gases. Since you took the ideal gas law and shaved it into your dog's fur, you'll no doubt remember that it goes like this:

$$PV = nRT$$

We said that we can describe a gas in terms of four main properties: mass, pressure, volume and temperature. If we have any three of these four, we can calculate the fourth easily. Here's one to chew on.

Calculate the temperature of 3.00 moles of argon in a 20.00 liter container at 2.00 atm pressure.

To start off, we can rearrange the ideal gas law equation in terms of temperature.

$$PV = nRT$$

$$T = \frac{PV}{nR}$$

That clears things up--now it's just a matter of plugging in the values we were given in the problem.

$$T = \frac{(2.00 \text{ atm} \times 20.00 \text{ liters})}{(3.00 \text{ mol} \times 0.0821 \text{ L-atm/mol K})} = 162 \text{ K}$$

Our answer is 162 Kelvins. Super duper, troopers.

Let's take a closer look at the ideal gas equation. Given that the following is true for n, we can calculate the molecular weight of an unknown gas.

$$n = \frac{\text{grams}}{\text{mole}} = \frac{\text{mass}}{\text{molecular weight}}$$

The ideal gas equation can be written as the following.

$$PV = \frac{\text{mass}}{\text{molecular weight RT}}$$

We can then rearrange the equation to read:

$$\text{molecular weight} = \frac{(\text{mass} \times R \times T)}{(PV)}$$

IS

With this equation, we can calculate the molecular weight of an unknown gas. Let's give it a whirl.

Calculate the molecular weight of an unknown gas when 0.970 g of the gas occupies 0.200 L at 99.0°C and 733 mm Hg.

We use the above equation and convert the mm Hg to atm.

$$\text{molecular weight} = \frac{(0.970g \times 0.0821 \text{ L-atm/mole K} \times 372 \text{ K})}{(0.200 \text{ L} \times 733mm/760mm/atm)}$$

$$= 154 \text{ g/mole}$$

We can also rearrange the ideal gas equation to give us an expression for the volume of a gas, which we'll need from time to time. The ideal gas equation is formulated for standard conditions of pressure and temperature (STP). These conditions are exactly 1 mole of substance, 1 atm of pressure, and 273.16 K.

$$PV = nRT$$

$$V = \frac{nRT}{P}$$

IMPORTANT STUFF

The Super-Charged World of Chemistry Parts 1, 2 & 3

At STP, this volume is called the molar volume and can be calculated for any gas.

$$V = \frac{(1.0 \text{ mole} \times 0.0821 \text{ L-atm/mol K} \times 273.16 \text{ K})}{1.00 \text{ atm}} = 22.4 \text{ L.}$$

Thus we've shown that at STP, 1 mole of an ideal gas will occupy 22.4 L.

> Farley, farley, farley, farley, farley, farley--Hafar!
>
> – Steve Martin in *The Three Amigos*

STUFF Nº. 7: PROPERTIES OF SOLUTIONS

Here's a typical problem that a chemist might encounter. It asks for the concentration of a solution, but only gives the density and the percent composition of the solution.

Calculate the molarity, molality, and mole fraction of a hydrochloric acid (HCl) solution whose density is 1.19 g/ml and whose percent composition is 37.2% HCl.

How do we go about this? Here's the battle plan. To determine molarity, we need to know the number of moles of HCl per liter of solution; to determine molality we need to know the number of moles of HCl per kilogram of solvent; and to determine the mole fraction, we need the moles of HCl and the number of moles of solvent (H_2O).

We'll start with the number of moles of HCl. For simplicity's sake, we'll assume a convenient amount of solution: 1000.0 mls, which is 1.000 liter of solution. This is not a random value that we've pulled out of thin air—it's precisely the amount we need to calculate molarity.

Multiplying the volume of the solution by the density, we get the mass of our 1000.0 mls: 1190 g. Since we already know that our solution is 37.2% HCl, we multiply the mass of 1000.0 mls times the percent composition and get the mass of our sample: 443 g.

$$1190 \text{ g} \times 0.372 = 443 \text{ g}$$

To find the number of moles in the sample, we divide the mass of our solution by the formula weight of the compound:

Remember, all you've got to do is add up the weights of the constituent atoms in the compound to get the formula weight.

$$\text{\# of moles of HCl} = \frac{443 \text{ g}}{36.46 \text{ g/mol}} = 12.2 \text{ moles}$$

Referring back to the molarity formula, we plug in the values we just determined.

$$\text{molarity} = \frac{\text{moles HCl}}{\text{volume of solution}} \text{ or } M = \frac{12.2 \text{ mol}}{1.0000 \text{ L}} = 12.2 \text{ M}$$

Next up to bat is molality. To calculate molality, we need to know the number of moles of HCl and the kilograms of solvent. Conveniently, we just found the number of moles of HCl (12.2 moles) and we also determined that out of the 1190 g of solution, 443 g were HCl.

Logically, the number of grams of solvent will be the total number of grams minus the grams of solute.

1190 g − 443g = 747 g or 0.747 kilograms

Note that this value has 3 significant figures.

Now that we've got the values for the number of moles of solute (HCl) and the number of kilograms of solvent, we're ready to work out the molality of our solution.

$$\frac{\text{moles HCl}}{\text{kilograms of solvent}} = \frac{12.2 \text{ moles}}{0.747 \text{ kilograms}} = 16.3 \text{ m}$$

Beautiful. Almost done! A quick dip in that chunky salsa called mole fractions, and we'll be all set to crunch this one.

To calculate the mole fraction of the solution, we need to know the number of moles of HCl (12.2) and the number of moles of H_2O. We figure out the number of moles of H_2O in the same way we did for HCl.

$$\text{moles of } H_2O = \frac{\text{mass of } H_2O}{\text{molecular weight of } H_2O}$$

$$= \frac{747 \text{ g}}{18.0 \text{ g/mol}}$$

$$= 41.5 \text{ moles}$$

We can now calculate the mole fraction of HCl and H_2O. Remember that the mole fraction equals the moles of a substance divided by the total number of moles in the solution.

$$X_{HCl} = \frac{\text{moles HCl}}{(\text{moles HCl} + \text{moles } H_2O)}$$

$$X_{HCl} = \frac{12.2}{(12.2 + 41.5)} = \frac{12.2}{53.7} = 0.227$$

$$X_{H2O} = \frac{\text{moles } H_2O}{(\text{moles HCl} + \text{moles } H_2O)}$$

$$X_{H2O} = \frac{41.5}{(12.2 + 41.5)} = \frac{41.5}{53.7} = 0.773$$

STUFF Nº. 8: COLLIGATIVE PROPERTIES

Let's take a closer look at the relationship between a change in freezing point and the molality of a substance.

AHHH-SO-REFRESHING REMINDERS:

• Δt_f = change in freezing point

• K_f = freezing point constant (which can be found in the table in your chem book)

• m = molality = $\dfrac{\text{moles solute}}{\text{kilogram of solvent}}$

• moles = $\dfrac{\text{mass of solute}}{\text{molecular weight of the solute}}$

Thus if you have the mass of the solute, the mass of the solvent, and the change in freezing point, you can calculate the molecular weight of the solute. This is similar to what we did with the ideal gas equation.

WHY NOT PUT ALL THAT FANTASTIC KNOWLEDGE TO WORK.

1.50 g of an unknown substance is dissolved in 75.0 g of cyclohexane and the freezing point is found to be 2.70°C. If the freezing point of pure cyclohexane is 6.50°C and the K_f for cyclohexane is 2.70°C/m, calculate the molecular weight of the unknown material.

We need to do several things to solve this bugger.

(1) Determine the change in the freezing point, Δt_f.

(2) Calculate m from $\Delta t_f = K_f m$ (since we are given K_f).

(3) Calculate the molecular weight of the solute using this formula:

$$m = \frac{\text{mass of solute} \times \text{kilograms of solvent}}{\text{molecular weight of the solute}}$$

A speedy subtraction gives us the change in freezing point.

$$\Delta t_f = 6.50\ °C - 2.70\ °C = 3.80\ °C$$

Next we use our favorite formula for molality.

$$\Delta t_f = K_f m$$

$$m = \frac{\Delta t_f}{K_f} = \frac{3.80}{2.70} = 1.41$$

Rearranging the following equation, we can work out the molecular weight of the solute.

$$m = \frac{\text{mass of solute}}{\text{molecular weight of the solute}} \times \text{kilograms of solvent}$$

$$\text{molecular weight} = \frac{\text{mass of the solute}}{(\text{mass of the solvent} \times \text{molality})}$$

$$= 1.50\text{g} \times \frac{1000 \text{ g}}{(1.41 \times 75)}$$

$$= 14.2 \text{ g/mol}$$

YOU CAN USE THIS SAME PROCEDURE FOR BOILING-POINT ELEVATION.

KEEP READING FOR MORE OTHER IMPORTANT STUFF IN OUR BASKET OF MOLECULES, MADNESS, AND MERRYMAKING!

STUFF №. 9: IONIC SOLUTIONS

You remember that colligative properties are dependent on the number of particles in solution. Until now, we have been discussing solutions that have a non-volatile solute that won't dissociate when dissolved in solution. For instance, if a non-volatile substance like sugar dissolves in water, for every one sugar molecule that dissolves, one sugar particle ends up in solution. However, when an ionic compound like NaCl dissolves in water, two particles result. This happens because NaCl is a strong electrolyte and breaks up into Na+ and Cl− ions.

Both the equations for the freezing-point depression and boiling-point elevation need to be modified to include the dissociation of the NaCl ions. So we do a little construction on the freezing-point formula to account for the number of ions formed when the ionic compound dissociates.

$$\Delta t_f = i K_f m$$

The **i** in this equation represents the number of ions formed when the ionic compound dissociates in water. The variable **i** is called the van't Hoff factor. It corrects the molality; $i \times m$ indicates the concentration of the total number of particles in the solution.

A few examples to whet your appetite:

For NaCl, $i = 2$

For $CaCl_2$, $i = 3$

STUFF Nº 10: PERIODIC TABLE

							18
							2 **He** HELIUM 4.00
		13	**14**	**15**	**16**	**17**	
			5 **B** BORON 10.81	6 **C** CARBON 12.01	7 **N** NITROGEN 14.01	8 **O** OXYGEN 16.00	9 **F** FLUORINE 19.00	10 **Ne** NEON 20.18

10	**11**	**12**						
			13 **Al** ALUMINIUM 26.98	14 **Si** SILICON 28.09	15 **P** PHOSPHORUS 30.97	16 **S** SULFUR 32.06	17 **Cl** CHLORINE 35.45	18 **Ar** ARGON 39.95
28 **Ni** NICKEL 58.71	29 **Cu** COPPER 63.55	30 **Zn** ZINC 65.38	31 **Ga** GALLIUM 69.72	32 **Ge** GERMANIUM 72.59	33 **As** ARSENIC 74.92	34 **Se** SELENIUM 78.96	35 **Br** BROMINE 79.90	36 **Kr** KRYPTON 83.80
46 **Pd** PALLADIUM 106.42	47 **Ag** SILVER 107.87	48 **Cd** CADMIUM 112.41	49 **In** INDIUM 114.82	50 **Sn** TIN 118.69	51 **Sb** ANTIMONY 121.75	52 **Te** TELLURIUM 127.60	53 **I** IODINE 126.90	54 **Xe** XENON 131.29
78 **Pt** PLATINUM 195.09	70 **Au** GOLD 196.97	71 **Hg** MERCURY 200.59	81 **Tl** THALIUM 204.37	82 **Pb** LEAD 207.19	83 **Bi** BISMUTH 208.98	84 **Po** POLONIUM (210)	85 **At** ASTATINE (2100	86 **Rn** RADON (222)

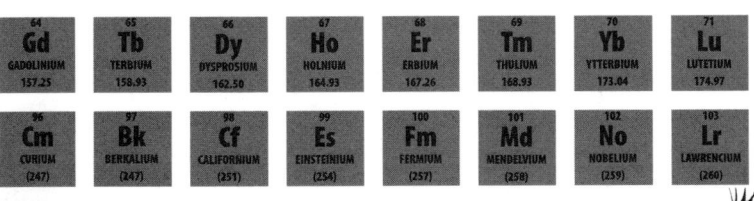

64 **Gd** GADOLINIUM 157.25	65 **Tb** TERBIUM 158.93	66 **Dy** DYSPROSIUM 162.50	67 **Ho** HOLMIUM 164.93	68 **Er** ERBIUM 167.26	69 **Tm** THULIUM 168.93	70 **Yb** YTTERBIUM 173.04	71 **Lu** LUTETIUM 174.97
96 **Cm** CURIUM (247)	97 **Bk** BERKALIUM (247)	98 **Cf** CALIFORNIUM (251)	99 **Es** EINSTEINIUM (254)	100 **Fm** FERMIUM (257)	101 **Md** MENDELVIUM (258)	102 **No** NOBELIUM (259)	103 **Lr** LAWRENCIUM (260)

IS

HERE AT ATOMIC MASS UNIVERSITY, A RENOWNED MOLECULAR UNIVERSITY, OUR DEVIANT STUDENT POPULATION COMPRISES THE RANGE OF THE PERIODIC TABLE. OUR STUDENTS ENJOY A RICH SOCIAL LIFE OF CHEMICAL INTERACTIONS WITH SIMILAR AND NON-SIMILAR ELEMENTS. STUDENTS EASILY BOND––IT IS COMMON TO SEE STUDENTS SHARING ELECTRONS AND FORMING COM-POUNDS ALL OVER CAMPUS.

DO YOU YEARN TO GET A TASTE OF AMU STUDENT LIFE? WITH AMU STUDENTLIFE PRODUCTS™ YOU, TOO, CAN BECOME ONE OF THE GANG! CHECK OUT OUR FALL LINEUP OF DESIGNER STUDENTLIFE PRODUCTS™.

These Tees are ELEMENTAL! Metals in parmesan, pesto, and peasant. Non-metals in barley, tofu, and cream of corn.

($39.95 XXL, XL, L, XS, XXS. Baby tees only $29.95)

The ultimate in collegiate coffeehouse wear, our wool Java-Java Jamocha Bean turtleneck is a steamin' concoction perfect for a long night at the cafe or an all-nighter spent huddled over an espresso and biscotti you've smuggled into the library. Posing was never meant to be this easy!

($99.95 one size fits all, black. Lightweight summer cotton available in latte, macchiato, and joe, $59.96)

Better than a blanket, our popular AMU sweatshirt is a favorite for all seasons. Our super high-tech fleece keeps you snuggly during those long ice hockey games, yet breathes enough to be used as a beach cover-up! Why not sport the athletic look while you're titrating in the lab?

($69.99 XXL. Comes in jock, jogger, and jumpshot.)

IS

Entire package only $34.95

This year we've inaugurated an entire line of AMU paper products. Tired of the same-old, same-old? AMU notebooks and lined paper come in electric yellow, chrome, and copper––and our AMU pens write in the very same colors!

(Decoder pen sold separately.)

Curl up with a steaming mug of hot chocolate or a cool draught of Mountain Dew--sound refreshing? Now you can work with an AMU mug at your side. Extra-broad base prevents spills and still leaves that time-honored coffee ring!

($24.95 in acid, base, and neutral)

Mystery. Seduction. Intermolecular forces you cannot deny. This stylish addition to your wardrobe will be the talk of your dorm! Experience ultimate comfort and haute couture in hatware. Go on, indulge in a truly luxurious coif! Why not keep several on hand to match your flannels or AMU sweat shirts?

We've got 'em in punk, ska, reggae, and acid jazz. ($29.95)

EXCLUSIVE!

For the more adventurous type: The Super Deluxe Space Helmet with Hyper Rotational Capacity. Because we all know what chem lab can be like——better to expect the unexpected! Ensure your own oxygen supply with this astro helmet, as seen in The Adventures of Orbital the Space Dog!

($999.95*)
* Ionizer capabilities sold separately.

The AMU bumpersticker lets everyone know where you learned how to balance equations——and how much you paid for that privilege! ($9.95 each)

A UFO? No, it's an AMU frisbee! It glows in the dark, whirs like a DC-9, and glides as sleekly as a goose heading south. Great for after-dark games in the chem lab!

($14.95 limited edition!)

IS

264

Parents, this is the wastebasket you've always dreamed of. Self-cleaning AMU trashcans come in extra large, really huge, and elephantine to accommodate your budding scholar's refuse. Fire engine red, pearly police car white, or elementary school bus yellow.

($29.95)

To thank you for your patronage and continued support of our chemical endeavors, we are proud to offer you your choice of these three deluxe-style, nouveau keychains when you purchase over $200 of AMU Studentlife Products™.

#1 is made of sturdy, weather-beaten steel clamps, melted down and burnished with a fine coating of gold dust and plaster of paris. Good luck losing these babies! (We'll be glad to engrave your keychain with your year of graduation.)

#2 is our hand-filigreed pendant of unusual delicacy. Contains enough radiation to read by at night.

#3 is constructed of spare nuclear reactor parts––well, heck, they are spare nuclear reactor parts! You can whip this keychain out when your attention wanders in class and just watch those little bolts of light bounce back and forth! (AMU is not responsible for zapped or singed materials.)

TEST YOURSELF

The Super-Charged World of Chemistry Parts 1, 2 & 3

PRACTICE EXAM 1

1. Define the following terms.

    a. homogeneous solution   _____

    b. limiting reagent   _____

    c. ion   _____

    d. mole   _____

    e. element   _____

    f. percent yield   _____

2. a. Balance this equation: $P_4 + C_{12} \rightarrow PCl_5$

    b. If 0.231 moles of P_4 are completely reacted according to the balanced chemical equation, how many moles of PCl_5 are produced?

3. How many *moles* of dioxane ($C_4H_8O_2$) are present in 5.80 g of dioxane? How many *molecules* are present in 5.80 g of dioxane?

TY

4. Name the following compounds.

    a. N_2O_3 _____

    b. HF   _____

    c. $Fe(ClO_4)_3$ _____

    d. $Ca_3\,(PO_4)_2$ _____

    e. HNO_2 _____

    f. NI_3 _____

    g. C_3PO

5. Borazine is a compound containing 40.31% boron (B), 52.18% nitrogen (N), and 7.51% hydrogen (H). What is the empirical formula for borazine? If its molar mass is 80 grams per mole, what is its molecular formula?

6. The density of gasoline is 0.7025 g/mL at 20°C. What is the mass of one gallon of gasoline? (1qt = 0.9463 L, 4 qt = 1 gal.)

TEST YOURSELF

The Super-Charged World of Chemistry Parts 1, 2 & 3

267

7. Given the following equation: $CS_2 + O_2 \rightarrow CO_2 + SO_2$

    a. What is the balanced chemical equation?

    b. If 10.00g of CS_2 reacts with 15.00 g of O_2, what mass of SO_2 will be produced?

8. You have 5.91 grams of a mysterious compound containing 2.03×10^{22} atoms. What is this compound's molar mass?

I can sing while I read,

I am singing––and reading––both!

– *Broadcast News*

PRACTICE EXAM 2

1. How many significant figures are in a quantity of mass measured as 0.05010 g?

   a. 1   **b.** 2   **c.** 3   **d.** 4   **e.** 5

2. Which of these are types of substances?

   a. compounds and homogeneous solutions

   b. compounds and heterogeneous solutions

   c. compounds and elements

   d. elements and homogeneous solutions

   e. elements and heterogeneous solutions

3. Which of these is an example of a chemical change?

   a. water boiling

   b. ice melting

   c. alcohol evaporating

   d. iodine vaporizing

   e. a sunburn

The Super-Charged World of Chemistry Parts 1, 2 & 3

4. In the depths of the chem lab, we cause the decomposition of a pure solid and--poof--we obtain a solid and a gas. From this information we can conclude with certainty that:

   a. the original solid is not an element

   b. both products are elements

   c. the second solid is a compound and the gas is an element

   d. the second solid is an element and the gas is a compound

5. The systematic name for BaH_2 is

   a. barium dihydrogen

   b. barium dihydride

   c. barium (II) hydrate

   d. barium dihydrate

   e. barium hydride

I'm picking out a thermos for you.

Not an ordinary thermos for you.

– *The Jerk*

6. The equation below shows the treatment of sodium borohy-
dride with sulfuric acid, which is a convenient method for the
preparation of diborane. When the equation is balanced, what
is the coeYcient for hydrogen?

____ $NaBH_4$ + ____ H_2SO_4 → ____ B_2H_6 + ____ H_2 + ____ Na_2SO_4

a. 1

b. 2

c. 3

d. 4

e. 5

7. Cumene is a compound composed of only C and H. It contains
89.9% C. What is its empirical formula?

a. CH   b. C_2H_3 c. C_3H_4 d. C_5H_9 e. C_7H_{10}

8. Calculate the number of moles of O_2 you need if you want it to
react with phosphorus to produce 5.50 g of P_4O_6. (Molecular
mass of P_4O_6 = 219.9 g/mol.)

a. 0.0250   b. 0.0500   c. 0.075   d. 0.150   e. 0.300

TEST YOURSELF

The Super-Charged World of Chemistry Parts 1, 2 & 3

9. If 50.0 g of O_2 are mixed with 50.0 g of H_2 and the mixture is ignited, what mass of water is produced?

    a. 50.0g   **b.** 56.3g   **c.** 65.7g   **d.** 71.4g   **e.** 100.0g

10. How much 0.54 M NaCl can be prepared via the dilution of 100 mL of a 6.0M NaCl solution?

    a. 1.1 L   **b.** 910 mL   **c.** 90 mL   **d.** 540 mL   **e.** 1.9 L

11. What is the formula for aluminum sulfite?

    a. $Al_2(SO_3)_3$

    b. $Al_2(SO_4)_3$

    c. $Al_3(SO_4)_2$

    d. Al_2S_3

    e. Al_3S_2

12. Main kindhearted dentists use a painkiller called procanime ($C_{13}H_{29}N_2O_2$). How many atoms of carbon are there in 0.80 moles of procanime?

    a. 4.8×10^{23}

    b. 9.6×10^{23}

    c. 3.2×10^{24}

    d. 6.3×10^{24}

    e. 9.6×10^{24}

13. Sulfur trioxide, SO_3, is made from the oxidation of SO_2. The reaction is represented by the equation $2SO_2 + O_2 \rightarrow 2SO_3$.

A 16 g sample of SO_2 gives 18 g of SO_3. What is the percent yield?

    a. 60%

    b. 75%

    c. 80%

    d. 90%

    e. 100%

14. The mass of fumaric acid, which occurs in many plants, is 41.4% carbon, 3.47% hydrogen, and 55.1% oxygen. A 0.050 M sample of this compound weighs 5.80 grams. What is the molecular formula of fumaric acid?

> Shwing!
>
> – *Wayne's World*

15. You're in salty Salt Lake City, Utah. You hop in your dune buggy and zoom off to San Francisco, which is 750 miles away. How far do you drive in kilometers? (1 in = 2.54 cm; 1 mile = 5,280 feet; 1 foot = 12 in.; 1 m = 100 cm; 1 km = 1000 m.)

Practice Exam 3

1. The smaller the difference in electronegativity between two atoms,

    1. the more ionic the bond

    2. the more covalent the bond

    3. the more polar the bond

        a. 1 only

        b. 2 only

        c. 3 only

        d. 1 and 3 only

        e. 2 and 3 only

2. What is the formal charge on the sulfur atom in sulfur trioxide (SO_3)?

    a. -2

    b. 0

    c. $+2$

    d. $+4$

    e. $+6$

3. The maximum number of electrons that can occupy the 5f orbitals is

   a. 5

   b. 7

   c. 10

   d. 14

   e. 18

4. Let's say you're examining the outermost electrons in a ground-state radium atom. Which of the following sets of values for the four quantum numbers (n, l, m_l, and m_s) would you use to describe the electrons?

   a. 6, 1, 1, ½

   b. 7, 0, 1, ½

   c. 7, 0, 0, −½

   d. 7, 1, 0, ½

   e. 7, 2, 1, −½

5. In which of these compounds does the central atom violate the octet rule?

    a. PF_3

    b. SiF_4

    c. OF_2

    d. ClF_3

    e. ClF

6. Define the following terms.

    a. standard state

    b. resonance structure

7. Write Lewis structures for the following molecules.

    a. XeF_4

    b. C_2H_2

    c. $BeCl_2$

    d. $NOCl$

8. What are the formal charges on the atoms in the following molecules?

    a. CO

    b. N_2O

    c. XeO_3

9. What do each of the following symbols refer to?

    a. n

    b. l

    c. m_l

    d. m_s

10. We can compare the shape of the p orbital to Orbital Woman holding a dumbbell in her hand. In this comparison, what part of the atom does the dumbbell shape represent?

11. How many electrons can be described by the following combination of quantum numbers?

    $n = 3$

    $l = 1$

    $ml = 1$

    $ms = +\frac{1}{2}$

TY

PRACTICE EXAM 4

1. Which of the following species of oxygen has a bond order of 1?

   1. O_2^-

   2. O_2^{2-}

   3. O_2^{2+}

     a. 1 only

     b. 2 only

     c. 3 only

     d. 1 and 2 only

     e. 2 and 3 only

2. Which of the following is the best explanation for a covalent bond?

     a. a chemical bond between two atoms that share a pair of electrons

     b. the overlapping of unoccupied orbitals of two or more atoms

     c. a positive ion attracting a negative ion

     d. an interaction between outer electrons

TEST YOURSELF

The Super-Charged World of Chemistry Parts 1, 2 & 3

3. An 8.1% sugar solution has a density of 1.05 g/cm³. What is the best value for the mass of sugar in 65.0 mls of this solution?

    a. 5.3 g

    b. 5.27 g

    c. 5.5 g

    d. 5.53 g

    e. 63 g

4. What would the van't Hoff factor (i) be for a 0.001 m solution of $K_2Cr_2O_7$?

    a. 1

    b. 2

    c. 3

    d. tr₄

    e. 5

5. Which of these properties is not a colligative property?

    a. freezing-point depression

    b. boiling-point elevation

    c. solubility

    d. osmosis

    e. vapor-pressure depression

6. What is the freezing point of a 0.25 m solution of glucose in water? (K_f for water is 1.86 °C/m)

    a. 0.93 °C

    b. 20.93 °C

    c. ⁻0.46 °C

    d. 20.46 °C

    e. 0.23 °C

7. Which of the following is a weak electrolyte in an aqueous solution?

    a. H_2S

    b. H_2SO_4

    c. HI

    d. HNO_3

    e. HCl

8. Your professor asks you to prepare 2.00 L of 0.100 M Na_2CO_3, and tells you that the molecular mass of Na_2CO_3 is 106. What is the best way for you to go about this experiment?

    a. Weigh out 21.2 g of Na_2CO_3 and add water until the final solution is homogeneous and has a volume of 2.00 L.

b. Weigh out 10.6 g of Na_2CO_3 and add water until the final solution is homogeneous and has a volume of 2.00 L.

c. Weigh out 21.2 g of Na_2CO_3 and add 2.00 kg of water.

d. Weigh out 10.6 g of Na_2CO_3 and add 2.00 kg of water.

e. Weigh out 21.2 g of Na_2CO_3 and add 2.00 L of water.

9. Imagine you are an alchemist's assistant. You stumble on the formula for turning gold into pencil lead and then also discover...the Ideal Gas! If you had to describe this gas, which pairs of variables would be directly proportional to each other (assuming that all other factors remain constant)?

   1. P, T

   2. P, V

   3. V, T

   4. n, V

      a. 1 and 2 only

      b. 3 and 4 only

      c. 2 only

      d. 1 and 3 only

      e. 1, 3, and 4 only

10. Which statement is inconsistent with the kinetic theory of an ideal gas?

 a. The forces of repulsion between gas molecules are very weak or negligible.

 b. Most of the volume occupied by a gas is empty space.

 c. When two gas molecules collide, they both gain kinetic energy.

 d. The average kinetic energy is proportional to the absolute temperature.

 e. Gas molecules move in a straight line between collisions.

11. The electron pairs in the XeF_4 molecule are arranged around the central atom. What shape do they form?

 a. a tetrahedron

 b. a trigonal bipyramid

 c. a square plane

 d. an octahedron

 e. a trigonal pyramid

TEST YOURSELF

The Super-Charged World of Chemistry Parts 1, 2 & 3

12. Neon atoms do not combine to form Ne$_2$ molecules, yet neon atoms can be liquefied as a result of which of the following intermolecular forces?

    a. dipole-dipole forces

    b. ion-dipole forces

    c. dipole-induced dipole forces

    d. dispersion forces

    e. nonmetal-nonmetal forces

13. Real gases deviate from ideal behavior because of the attractive forces between the gas molecules and

    a. ionization energies

    b. pressures within the chemical bonds

    c. the molecules as having different velocities

    d. the actual volume of the molecules

    e. the forces of the Dark Side

TY

14. A container holds 2.0 mol of oxygen, 3.0 mol of nitrogen, and 1.0 mol of carbon dioxide. What is the partial pressure of nitrogen if the total pressure in the container is 900 mm Hg?

    a. 150 mm Hg

    b. 200 mm Hg

    c. 300 mm Hg

    d. 450 mm Hg

    e. 600 mm Hg

15. Which species has the largest number of pairs of electrons around its central atom?

    a. $XeOF_4$

    b. XeF_6

    c. XeF_4

    d. XeF_2

    e. SiF_6^{2-}

The Super-Charged World of Chemistry Parts 1, 2 & 3

16. Calcium nitrate will react with ammonium chloride at slightly elevated temperatures, as represented in this equation:

$$Ca(NO_3)_2(s) + 2NH_4Cl(s) \rightarrow 2N_2O(g) + CaCl_2(s) + 4H_2O(g)$$

A 2.00 mol sample of each reactant will give what volume of N_2O at STP?

   a. 11.2 L

   b. 22.4 L

   c. 33.6 L

   d. 44.8 L

   e. 67.2 L

17. Which of the following indicates the existence of strong inter-molecular forces of attraction in a liquid?

   a. a very low boiling point

   b. a very low vapor pressure

   c. a very low critical pressure

   d. a very low viscosity

   e. a very low heat of vaporization

   f. a very low rate of socialization

18. What is the molecular orbital diagram for He_2^+? Do you expect the ion to be stable? Explain.

19. Some curious chemists analyzed a hydrocarbon and found that it was 14.4% hydrogen and 85.6% carbon by weight. If the hydrocarbon compound has a molecular mass of 138 g/mole, what is its molecular formula?

The only kind of love is stone-blind love.

– Tom Waits

PRACTICE EXAM 5

(OTHER IMPORTANT STUFF)

1. Let's say your favorite aunt gives you an awesome computer game, Capture the Planet, and a compound containing only C and H. The compound has a percent composition of 92.30% C and 7.70% H.

   a. If the molar mass of the compound is 78 g/mol, what is its molecular formula?

   b. If you spend four hours a day playing your new computer game, how long will it take you to figure out how to rule the planet?

2. Acetone is a compound that contains C, H, and O, and has a percent composition of 62.0% C, 10.4% H and 27.5% O. If its molar mass is 58 g/mol, what is the molecular formula of acetone?

3. While rummaging through a drawer, you find an organic compound. You analyze it and discover it contains 58.8% C, 9.80% H, and 31.4% O, and has a molar mass of 204 g/mol. What are the empirical and molecular formulas of the compound?

4. Name the following compounds.

    a. K_2SO_4

    b. Cr_2O_3

    c. $FeCl_3$

    d. Na_2SO_3

    e. $Ca(CN)_2$

    f. H_2S

    g. N_2O_5

    h. N_2O

    i. P_2O_5

    j. $Co(NO_3)_3$

    k. $Mn(NO_2)_2$

    l. ClF_3

5. Write the formulas for the following compounds.

    a. sodium acetate

    b. calcium oxalate

    c. chloric acid

    d. oxygen difluoride

    e. tin (II) bromide

    f. phosphorus pentafluoride

    g. dichlorine monoxide

    h. iron (II) sulfate

TEST YOURSELF

The Super-Charged World of Chemistry Parts 1, 2 & 3

6. Determine **i** for these compounds.

    a. NaI

    b. $AlCl_3$

    c. $MgSO_4$

    d. $Mg(NO_3)_2$

    e. Sugar

7. What is the geometry of these compounds?

    a. XeF_4

    b. IF_5

    c. ICl_3

    d. XeF_2

8. For each of the compounds in question 7, determine the hybridization around the central atom.

9. What is the volume of 0.25 moles of nitrogen (N_2) at STP?

10. Calculate the molecular weight of Mystery Gas if 0.427 grams of the gas occupies 185 ml at 755 mm Hg and 100.0 °C.

TY

11. An expert chemist prepares a solution by placing 0.131 g of a substance in 25.4 g of water. If the freezing point of the solution is −0.104 °C, what is the molecular weight of the material she has created? (K_f = 1.86 K kg/mol for water)

12. Calculate the molarity, molality, and mole fraction for these two aqueous solutions.

    a. An $HClO_4$ solution which has a density of 1.54 g/ml and is 60.0% $HClO_4$ by weight.

    b. A KOH solution with a density of 1.46 g/ml and 45.0% KOH by weight.

THE ANSWER MAN

The Answer Man is dedicated to alleviating the worries of his fellow Cerebellum employees. Each day, he receives hundreds of letters from co-workers. We have reprinted some of them here with The Answer Man's responses. We hope they can be helpful to you, too.

Dear Answer Man:

Why do we drive on parkways and park on driveways? If potatoes grow eyes and corn grows ears, what kind of vegetable grows noses?

And finally...

When is the world going to end?

How odd that you should ask all three of these questions together. The world will end after 6,000 more uses of those first two clichés. Thank you for edging us ever so much closer to that eventuality.

Dear Answer Man:

I have a nagging suspicion that the world is speeding through space and there is nobody at the control console. Do I have reason to be worried?

No, no. We're not speeding.

Dear Answer Man:

Do the lights go out when we close our eyes?

Yes. There is a little elf called Reekbo who jumps out from hiding and turns off the lights when you close your eyes. He waits for you to open them again, and turns the lights back on right before you do. Then Reekbo hides again. He's very fast.

Dear Answer Man:

I have been having a recurring dream in which I am a bumble-bee buzzing around through the air, going from flower to flower in a meadow, and the flowers are really weird, like all the petals are different shapes and sizes, and I can smell the pollen, and I think to myself, I'm allergic to pollen, I can't be a bumblebee! And then all of the sudden I wake up. Why?

Obviously, because you're allergic to pollen. If you sneezed while in bee form, you'd blow your wings off.

STRESS RELIEF

The Super-Charged World of Chemistry Parts 1, 2 & 3

CHEMCROSS BRAIN CRUNCHER

ACROSS

1. takes up space and has mass
3. properties you can only find by experimenting
7. Z, the atomic...
8. positive particle
11. Helium Man is this kind of gas
12. moles of solute in one liter of solution
14. answers the question "how much?"
16. chemistry rodent
17. the yield you really get
19. solid to liquid is a change of...
21. 6 C, 12 H, 6 O
23. Zn
25. adding a solvent
30. combustible experiments some-times...
32. like neutrons and Switzerland
34. New York City alligators swim in...
35. monkey with vividly colored buttocks
36. measures against "true value"

DOWN

2. can't be broken down chemically
4. atom with different number of neutrons
5. the chemistry wiz who invented moles
6. F
9. property such as melting point
10. when acids equal bases
13. the same throughout
15. made up of different stuff
18. natural amount of an isotope
20. Np
22. this formula gives exact numbers
24. charged-up atom
26. gouda's stinky cousin
27. ≥ 2 kinds of atoms combined
28. the noble superhero element
28. Granny Smith or horse
31. dribble
33. mole impersonator

294 CHEMSEARCHO

```
F E R M I U M A N G A N E S E M
P H O S P H O R U S O D E D P M
M Y M H N L L D N D A L L O U T
U D N A E T Y U A E A O T I I E
N R O F G A B R L M G A N L I E
I O S N M N D C U U S O I O T F
M G E I T G E I H S C M R U E L
U E N U Ⓐ C N S I R B N N T A N
L N I M Ⓒ O U U I U O G I C I E
A N R H Ⓣ P M Z R U S M I Z E N
R O O U Ⓘ P E G S T M M I I D I
E T L S Ⓝ E E Z E W E E D U O R
V P H K Ⓘ R O N R H E L I U M O
L Y C R Ⓤ A N A C R A D I U M U
I R T O Ⓜ R T I T A N I U M O L
S K I W I S U P E R H E R O C F
```

SOLUTION: 19 LETTERS

Look through the scrambled letters for the words listed to the right. Circle each letter of each word as you find it. When you circle all of the letters of all the words, some letters will be left over. Unscramble the unused letters to solve the puzzle.

actinium, aluminum, chemical, chlorine, chromium, commode, copper, crud, feet, fermium, fluorine, gnat, gold, hafnium, helium, hydrogen, iron, kiwi, krypton, lead, limburger, magnesium, manganese, mole, molybdenum, neon, nitrogen, nose, phosphorus, plutonium, potassium, radium, radon, rare, silver, sneezeweed, superhero, tact, titanium, tungsten, work, zinc, zirconium

CHEM SAFETY GUY IS
ON YOUR SIDE.

A

Quiz 1

1. a. mixture

    b. compound

    c. compound

    d. element

    e. mixture

    f. mixture

    g. mixture

    h. element

    i. mixture

2. a. 5.40

    b. 19.2

    c. 2.5×10^2

    d. 3.0×10^4

    e. 6.000

    f. 1.243×10^5

The Super-Charged World of Chemistry Parts 1, 2 & 3

3. a. milliliters $= 2.07 \text{ gal} \times \dfrac{4 \text{ qt}}{1 \text{ gal}} \times \dfrac{0.9464 \text{ L}}{1 \text{ qt}} \times \dfrac{1000 \text{ ml}}{1 \text{ L}}$

$= 7.84 \times 10^{3} \text{ mL}$

b. miles $= 1.07 \times 10^{5} \text{ cm} \times \dfrac{1 \text{ in}}{2.54 \text{ cm}} \times \dfrac{1 \text{ ft}}{12 \text{ in}} \times \dfrac{1 \text{ mile}}{5280 \text{ ft}}$

$= 6.65 \times 10^{-1} \text{ miles}$

c. lbs $= 8.72 \times 10^{-9} \text{ grams} \times \dfrac{1 \text{ kg}}{1000 \text{ g}} \times \dfrac{1 \text{ lb}}{0.4536 \text{ kg}}$

$= 1.92 \times 10^{-11} \text{ lbs}$

He'll put your head out a puppet
on a string paid to sing or rhyme or
do my thing I'm in a lava lamp
inside the brain hotel I might be
freakin' and peakin' but I rock
well...

– The Beastie Boys

A

QUIZ 2

1.   a. $C_5H_{12} + 8O_2 \rightarrow 5CO_2 + 6H_2O$

       b. $AlCl_3 + 3H_2O \rightarrow Al(OH)_3 + 3HCl$

       c. $Cl_2O_7 + H_2O \rightarrow 2HClO_4$

       d. $3NO_2 + H_2O \rightarrow 2HNO_3 + NO$

       e. $Fe_3O_4 + 4H_2 \rightarrow 3Fe + 4H_2O$

2.   Average atomic weight of chlorine

       $= 34.96885 \times 0.75771 + 36.96590 \times 0.24229$

       $= 35.453$ g/mol.

3.   a. 78.11 g/mol

       b. 108.010 g/mol

       c. 283.889 g/mol

4. n (number of moles) = mass/formula mass (FM);
     FM = mass/n = 7.82 g/0.139 mol = 56.3 g/mol

5. a. %S = 40.0%;

%O = 60.0%

b. %C = 32.2%;

%N = 15.3%;

%O = 52.5%

c. %N = 30.4%;

%O = 69.6%

6. Assuming 100.0 grams of material:

mol C = 4.10; mol H = 6.85; mol O = 2.74.

ratio of atoms = $C_{1.50}H_{2.50}O_1$ or $C_6H_{10}O_4$

7. $4Fe + 3O_2 \rightarrow 2Fe_2O_3$:

$$\text{moles of } Fe_2O_3 = 0.150 \text{ mol } O_2 \times \frac{2 \text{ mol } Fe_2O_3}{3 \text{ mol } O_2}$$

$$= 0.100 \text{ mol } Fe_2O_3$$

$$\text{grams of } Fe_2O_3 = 0.100 \text{ mol } Fe_2O_3 \times 159.7 g/mol$$

$$= 16.0 \text{ g } Fe_2O_3$$

8. $2N_2O_5 \rightarrow 4NO_2 + O_2$

moles $N_2O_5 = 2.427 \times 10^{-2}$

moles $NO_2 = 4.854 \times 10^{-2}$

grams $NO_2 = 2.233$

ANSWERS

The Super-Charged World of Chemistry Parts 1, 2 & 3

QUIZ 3

1. a. $M = \dfrac{n}{V}$

    $n = 2.93 \times 10^{-2}$ mol
    molar concentration of $KNO_3 = 0.117$ M

    b. $M = \dfrac{n}{V} = 0.210$ M

2. $V_c = 0.0313$ L or 31.3 mL of 12.0 M HCl

3. $H_2SO_4 + 2NaOH \rightarrow Na_2SO_4 + 2H_2O$

    $V = \dfrac{n}{M}$

    $n_{NaOH} = 2 \times n_{H2SO4} = 4.73 \times 10^{-3}$ mol

    $V = \dfrac{4.73 \times 10^{-3} \text{ mol NaOH}}{0.100 \text{M NaOH}}$

    $= 4.73 \times 10^{-2}$ L or 47.3 mL

A

302

4. $4Al + 3O_2 \rightarrow 2Al_2O_3$

    limiting reagent is Al

    moles of $Al_2O_3 = 0.0500$ mol

5. $6Cl_2 + P_4 \rightarrow 4PCl_3$

    limiting reagent is Cl_2

    $$\text{moles of } PCl_3 = 0.296 \text{ mol } Cl_2 \times \frac{4 \text{ mol } PCl_3}{6 \text{ mol } Cl_2} \times 137.3 \text{ g/mol } PCl_3$$

    $$= 27.1 \text{ g } PCl_3$$

    $$\text{remaining } P_4 = 10.429 - 0.296 \times \frac{1 \text{ mol } P_4}{6 \text{ mol } Cl_2} \times 123.9 \text{ g/mol } P_4$$

    $$= 47.0 \text{ g } P_4$$

6. $$\% \text{ yield} = \frac{\text{actual yield}}{\text{therotical yield}} \times 100$$

    $$= \frac{19.5 \text{ g } C_6H_{12}NO_2}{29.0 \text{ g } C_6H_{12}NO_2} \times 100$$

    $$= 67.2\%$$

The Super-Charged World of Chemistry Parts 1, 2 & 3

QUIZ 4

1. $q = 13.5g \times 0.385 \text{ J/(g }^\circ\text{C)} \times 0.398\ ^\circ\text{C} = 2.07\ \text{kJ}$

2. $\Delta H = 2\Delta H^\circ_{f\ CO_2}\ (g) + 3\Delta H^\circ_{f\ H_2O}\ (l) - \Delta H^\circ_{f\ C_2H_6}\ (g)$

   $= -2 \times 393\ \text{kJ} - 3 \times 285.8\ \text{kJ} + 84.7\ \text{kJ}$

   $= -1559\ \text{kJ}$

3. $\Delta H = \Delta H^\circ_{f\ CaSiO_3} - (\Delta H^\circ_{f\ CaO} + \Delta H^\circ_{f\ SiO_2})$

   $-89.5\ \text{kJ} = \Delta H^\circ_{f\ CaSiO} - (-635\ \text{kJ} - 911\ \text{kJ})$

   $\Delta H^\circ_{f\ CaSiO} = -1640\ \text{kJ}$

QUIZ 5

1. P has an *n* value of 3 and N has an *n* value of 2. Since the larger the *n* value, the further away the outermost electrons are from the nucleus, we can conclude that P is the element whose electrons are furthest away from its nucleus.

2. $n = 3, l = 0, m_l = 0$, and $m_s = \frac{1}{2}$

3. b, c, d

**His brain is as dry as the remainder
biscuit after a voyage.**

   – As You Like It

ANSWERS

QUIZ 6

1. a.

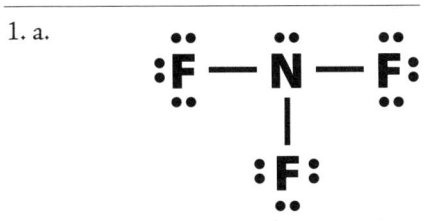

b.

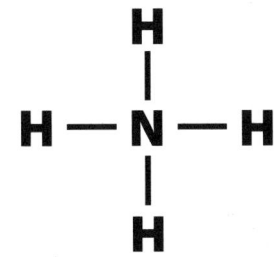

c.

$$H - N - H$$

with H above and H below the N

A

d.

$$:\ddot{Br} — \ddot{Br}:$$

e.

$$:N ≡ N:$$

f.

$$:\ddot{F}:$$
$$|$$
$$Al — \ddot{F}:$$
$$|$$
$$:\ddot{F}:$$

g.

$$[:N ≡ O:]^+$$

ANSWERS

The Super-Charged World of Chemistry Parts 1, 2 & 3

2. a. Electronegativity describes the ability of an atom in a molecule to attract the electrons of a covalent bond.

    b. Polar covalent bonds are covalent bonds between two atoms with very different electronegativities (the atoms don't share electrons equally).

    c. Ionic bonds are chemical bonds formed by the electrostatic attraction between positive and negative ions.

    d. Covalent bonds are chemical bonds that form when atoms share a pair of electrons.

    e. Bond energy is the energy required to break the bonds between the atoms in one mole of a substance.

3. For problems 3a through 3c, we used Ebbing values (check your textbook to see which set of values it uses). Also, for these problems,

    $$\Delta H^\circ = \Sigma \Delta H^\circ(\text{bonds broken}) - \Sigma \Delta H^\circ(\text{bonds formed})$$

    a. $\Delta H^\circ = (C-H + Br-Br) - (C-Br + H-Br)$
       $= -46kJ$

    b. $\Delta H^\circ = (4Cl-Cl + 4C-H) - (4C-Cl + 4H-Cl)$
       $= -416kJ$

    c. $\Delta H^\circ = (6C-H + C-C + \frac{7}{2}O_2) - (4CO + 6H-O)$
       $= -1671\ kJ$

A

QUIZ 7

1. a. linear

   b. trigonal planer

   c. pyramid

   d. bent planer

   e. tetrahedral

2. a. sp

   b. sp^2

   c. sp^3

   d. sp^3

   e. sp^3

3. a. An antibonding molecular orbital is an orbital that concentrates electrons in places other than between the nuclei of the two bonded atoms. Antibonding molecular orbitals have higher energy than bonding molecular orbitals.

   b. Bond order refers to the stability of a covalent molecule, and equals one-half the difference between the number of bonding electrons and the number of antibonding electrons.

   c. A chemical bond formed between two atoms that share a pair of electrons.

ANSWERS

The Super-Charged World of Chemistry Parts 1, 2 & 3

QUIZ 8

1. a. A gas is composed of molecules that are far apart from each other. Most of the volume a gas occupies is nothing but empty space.

    b. Gas molecules are in constant, random motion. Each molecule continues to move in a straight line unless it collides with another molecule or with a wall of the container.

    c. Gas molecules exert no force on each other or on the container, except when they collide with each other or with the walls of the container.

    d. The average kinetic energy of the molecules in a gas is proportional to the temperature of the gas.

    e. Every time a molecule collides with a wall, it exerts a force on the wall.

2. $PV = nRT$

    $$P = \frac{nRT}{V}$$

    $$n = \frac{7}{28} = 0.25$$

    $$P = 0.25 \times 0.0821 \times \frac{298}{1} = 6.1 \text{ atm}$$

A

3. $PV = nRT$

$$P = \frac{nRT}{V}$$

$$n = \frac{10.0}{20.2} = 0.495$$

$$\text{partial pressure} = 0.495 \times 0.0821 \times \frac{298}{1} = 12.1$$

$$\text{total pressure} = 12.1 + 6.1 = 18.2 \text{ atm}$$

Peace, good pintpot, peace, good
tickle-brain.

   – *Henry IV, Part I*

ANSWERS

The Super-Charged World of Chemistry Parts 1, 2 & 3

QUIZ 9

1. a. Dipole-dipole forces are intermolecular forces that result from the slight charges on the ends of polar molecules and cause the molecules to stick together.

   b. Hydrogen bonds are intermolecular bonds that exist between hydrogen atoms in one molecule and electronegative atoms such as oxygen or nitrogen in another molecule (or different part of the same molecule.)

   c. London dispersion forces are intermolecular forces that result from fluctuating, instantaneous dipoles.

   d. The triple point represents the pressure and temperature at which all three phases of a substance are at equilibrium.

   e. A phase change occurs when a substance converts from one state of matter to another.

   f. In a closed system, vapor pressure is the constant force that the vapor exerts above the liquid once equilibrium has been reached.

A

2.　a. melting

　　b. critical point

　　c. freezing

　　d. vaporization

　　e. condensation

　　f. sublimation

　　g. triple point

　　h. deposition

If his name be George, I'll call him Peter.

　– *King John*

QUIZ 10

1. a. Miscible liquids are liquids that can dissolve in each other.

    b. A saturated solution has accepted as much solute as it possibly can; any further solute added to the solution will remain undissolved.

    c. Molality (m) equals the number of moles of solute per kilogram of solvent.

    d. The mole fraction (X) of a compound is equal to the number of moles of the compound you're measuring, divided by the total number of moles of all components of the system.

    e. Molarity (M) is the number of moles of solute per liter of solution.

    f. Osmosis is the flow of solvent through a semipermeable membrane. The solvent flows from the more dilute solution to the more concentrated one.

2. mole fraction $= \dfrac{\text{moles KNO}_3}{\text{total moles}}$

moles KNO$_3$ $= \dfrac{31.2}{101.1} = 0.309$

moles H$_2$O $= \dfrac{100}{18} = 5.55$

$X_{\text{KNO}_3} = \dfrac{0.309}{5.86} = 0.0527$

$m = \dfrac{m}{1000} = \dfrac{0.309}{0.100} = 3.09$

3. $M = \dfrac{\text{mol}}{V} = \dfrac{0.309}{0.131} = 2.36$

4. $\Delta t_f = mK_f = 0.250 \times 1.86 = 0.465$

freezing point $= {}^-0.465$ °C

$\Delta t_b = mK_b = 0.250 \times 0.521 = 0.130$

boiling point $= 100.130$ °C

5. Take 130×0.0305 to get the mass of MgCl$_2$, which equals 3.97 grams. Weigh 3.97 grams of MgCl$_2$ and add it to 126 grams of water.

ANSWERS

The Super-Charged World of Chemistry Parts 1, 2 & 3

QUIZ XYZ

If all the people in the world were atoms, we'd fit on the head of a pin.

– Professor Chemster

1. b

2. c

3. all except for c

PRACTICE EXAM 1

1. a. A homogeneous solution is a mixture of substances that has the same physical and chemical properties throughout.

    b. In a reaction, the limiting reagent is the reactant that limits the amount of product formed.

    c. An ion is a positively or negatively charged atom.

    d. A mole is a unit of measurement that represents 6.022×10^{23} things.

    e. An element is a substance that cannot be broken down chemically into a simpler substance.

    f. The percent yield equals the actual yield divided by the theoretical yield, times 100.

2. a. $P_4 + 10Cl_2 \rightarrow 4PCl_5$

    b. moles of PCl_5 = 0.231 mol $P_4 \times \dfrac{4 \text{ mol } PCl_5}{1 \text{ mol } P_4}$

    = 0.924 mol PCl_5

ANSWERS

The Super-Charged World of Chemistry Parts 1, 2 & 3

3. # of moles = n = $\dfrac{5.80\text{g C}_4\text{H}_8\text{O}_2}{88.1 \text{ g/mol}}$

 = 6.58×10^{-2} mol $C_4H_8O_2$

 # of molecules = n $\times$ 6.02×10^{23}
 = 3.96×10^{23} molecules of $C_4H_8O_2$

4. a. dinitrogen trioxide

 b. hydrofluoric acid

 c. iron (III) perchlorate

 d. calcium phosphate

 e. nitrous acid

 f. nitrogen triodide

 g. protocol droid

[NOTE: FOR QUESTIONS 5, 6, AND 7, WE ROUNDED UP OUR VALUES.]

5. empirical formula $= BNH_2$

molecular formula $= (BNH_2)_x$

x = empirical formula\molar mass $= {}^{80}\!/_{27} = 3$

molecular formula $= B_3N_3H_6$

6. $\text{mass} = \dfrac{0.7025 \text{ g}}{\text{mL}} \times \dfrac{1000\text{mL}}{\text{L}} \times \dfrac{0.9463 \text{ L}}{\text{qt}} \times \dfrac{4 \text{ qt}}{\text{gal}} \times \dfrac{1 \text{ kg}}{1000 \text{ g}}$

$= 2.66$ kg/gal, the density of gasoline

7. a. $CS_2 + 3O_2 \rightarrow CO_2 + 2SO_2$

b. CS_2 is the limiting reagent

$\text{mass of } SO_2 = 0.132 \text{ mol } CS_2 \times \dfrac{2 \text{ mol } SO_2}{1 \text{ mol } CS_2} \times \dfrac{64.0 \text{ g } CS_2}{1 \text{ mol } CS_2}$

$= 16.9$ grams

ANSWERS

The Super-Charged World of Chemistry Parts 1, 2 & 3

8. $n = \dfrac{\text{mass}}{\text{MM}}$ or

   $MM = \dfrac{\text{mass}}{n}$

   $= 5.91 \text{ g} / (2.03 \times 10^{22} \text{ atoms} / 6.02 \times 10^{23}) \text{ atoms/mole}$

   $= 175 \text{ g/mol}$

PRACTICE EXAM 2

1. d	6. b.	11. a
2. c	7. c	12. d
3. e	8. c	13. d
4. a	9. b	
5. e	10. a	

14. empirical formula = CHO

molecular formula = $(CHO)_x$

$$x = \frac{\text{molar mass}}{\text{molecular weight}} = {}^{116}\!/_{29} = 4$$

molecular formula = $C_4H_4O_4$

15. kilometers $= 750 \text{ mile} \times \dfrac{5280 \text{ ft}}{\text{mile}} \times \dfrac{12 \text{ in}}{\text{ft}} \times \dfrac{2.54 \text{ cm}}{\text{in}} \times \dfrac{1\text{m}}{100\text{cm}} \times \dfrac{\text{km}}{1000\text{m}}$

$= 1207 \text{ km or } 1.21 \times 10^3 \text{ km}$

The Super-Charged World of Chemistry Parts 1, 2 & 3

PRACTICE EXAM 3

1. b

2. b

3. d

4. c

5. d

6. a. A substance is in its standard state when it is at 25°C, one atmosphere pressure, and in its most stable form.

    b. When there is more than one way to diagram the Lewis structure of a molecule, the possible diagrams are called resonance structures. Chemists and chemistry students alike take the "average" of a molecule's resonance structures to arrive at the accepted Lewis structure for that molecule.

7. a.

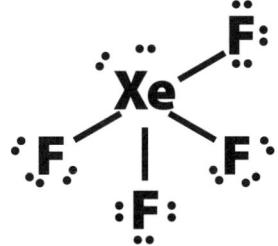

b.

$$H - C \equiv C - H$$

c.

$$:\ddot{C}l - Be - \ddot{C}l:$$

d.

$$:\ddot{C}l - \ddot{N} = \ddot{O}:$$

8. a.

$$:C \equiv O:$$

(-1) (+1)

b.

$$:N \equiv N \rightarrow \ddot{O}:$$

(+1) (-1)

c.

$$(-1) :\ddot{O} - \overset{(+3)}{\underset{|}{X}e} - O \ (-1)$$
$$:\ddot{O}:$$
$$(-1)$$

9. n relates to distance, l refers to distribution in space, m_l relates to orientation, and m_s refers to the electron spin.

10. Electron distribution around the nucleus.

11. One.

Chill pick your teeth.

– King Lear

A

PRACTICE EXAM 4

1. e	7. a	13. d
2. a	8. a	14. d
3. c	9. e	15. b
4. c	10. c	16. d
5. c	11. d	17. b
6. c	12. d	

18. σ_{1s}^* __ (↑) __ yes, bond order is ½

   σ_{1s} __ (↑↓) __

19. $\dfrac{14.4}{1.008} = 14.3$

   $\dfrac{85.6}{12.01} = 7.13$

   empirical formula = CH_2
   empirical weight = 14
   molecular formula = $(CH_2)x$
   $x = {}^{138}/_{14} = 10$
   formula = $C_{10}H_{20}$

ANSWERS

PRACTICE EXAM 5

(OTHER IMPORTANT STUFF)

1. a. Here are two different ways to solve this problem.

   Method # 1

   (1) Determine mass of each element in one mole of compound.

   (2) Use these as coefficients in molecular formula and you're done!

   mass of C = 0.9239 × 78 = 72 g of C

   mass of H = 0.0770 × 78 = 6 g of H

In one mole of compound there are 6 moles of C and 6 moles of H; so the molecular formula must be C_6H_6.

   Method # 2

   Empirical formula is CH. Molecular formula is $(CH)x$.

   $$x = \frac{\text{molar mass}}{\text{molecular weight}}$$

   $$x = \frac{78 \text{ g mole}}{13g} = 6$$

   molecular formula = C_6H_6

A

      b. Depends on your proficiency. Average person = one month; whiz = two weeks; cyberpunk = less than a week.

2.   empirical formula = C_3H_6O

      molecular formula = $(C_3H_6O)x$

      $x = {}^{58}/_{58} = 1$

      Molecular formula and empirical formula are the same.

(You may also use Method #1 as outlined in the answer for question #1.)

3.   empirical formula = $C_5H_{10}O_2$

      molecular formula = $(C_5H_{10}O_2)x$

      $x = {}^{204}/_{102} = 2$

      molecular formula = $C_{10}H_{20}O_4$

4.   a potassium sulfate

      b. chromium (III) oxide

      c. iron (III) chloride

      d. sodium sulfite

      e. calcium cyanide

      f. hydrogen sulfide

ANSWERS

The Super-Charged World of Chemistry Parts 1, 2 & 3

g. dinitrogen pentoxide

h. dinitrogen oxide

i. phosphorous pentoxide

j. cobalt (III) nitrate

k. manganese (II) nitrite

l. chlorine trifluoride

5. a. $NaC_2H_3O_2$

    b. CaC_2O4

    c. $HClO_3$

    d. OF_2

    e. $SnBr_2$

    f. PF_5

    g. Cl_2O

    h. $FeSO_4$

6. a. 2

    b. 4

    c. 2

    d. 3

    e. 1

A

7. a. square planer

    b. square pyramid

    c. T-shaped

    d. linear

8. a. d^2sp^3

    b. d^2sp^3

    c. dsp^3

    d. dsp^3

9. $0.25 \times 22.4 = 5.6$ L

10. molecular weight $=$ mass $\times \dfrac{RT}{PV}$

    $$\frac{(0.427 \times 0.0821 \times 373)}{(0.185 \times {}^{755}\!/_{760})} = 71.1 \text{ g/mol}$$

11. $\Delta t_f = K_f m$

    $m = \Delta t_f / K_f = 0.104/1.86 = 0.0559$

    $mw = \text{mass}/(\text{kg} \times m)$

    $\quad = 00.131/(0.0254 \times 0.0559)$

    $\quad = 92.3$ g/mol

12. a. 1000.0 mL of $HClO_4$ contains 924 g of $HClO_4$ and 616 g of H_2O

   moles $HClO_4 = {}^{924}/_{100} = 9.24$

   moles $H_2O = 34.2$

   $M \times {}^{9.24}/_{1.000} = 9.24$

   $m = {}^{9.24}/_{0.616} = 15.0$

   $X_{HClO_4} = {}^{34.2}/_{43.4} = 0.213$

   $X_{H_2O} = {}^{9.24}/_{43.4} = 0.788$

   b. Same procedure as above.

   Moles of KOH = 11.7

   moles of $H_2O = 44.6$

   $M = {}^{11.7}/_{1.000} = 1.7$

   $m = {}^{11.7}/_{0.803} = 14.6$

   $X_{KHO} = {}^{11.7}/_{56.3} = 0.208$

   $X_{H_2O} = {}^{44.6}/_{56.3} = 0.792$

> You Banbury cheese!
>
> – *The Merry Wives of Windsor*

A

CHEMCROSS BRAIN CRUNCHER SOLUTION

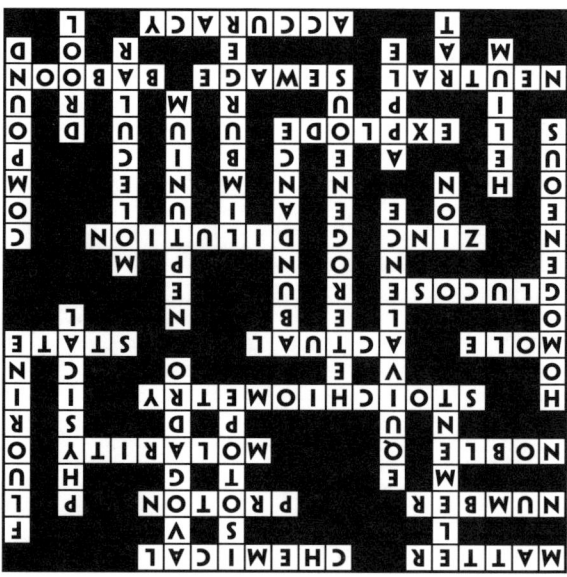

CHEMSEARCHO SOLUTION

Helium man is inert.

GLOSSARY

GLOSSARY

For your convenience and reading pleasure, we've bolded terms from the glossary the first time they appear in the text.

accuracy – Refers to how close a measurement is to the theoretical "true value."

acid – A substance containing hydrogen ions. The larger the quantity of hydrogen ions, the stronger the acid.

acid-base indicator – A weak acid or base that changes color when the amount of hydrogen ions in the solution reaches a certain value.

actual yield – The amount of the product actually produced in a reaction.

anion – An ion with a negative charge.

antibonding molecular orbital – A molecular orbital with higher energy and less stability than bonding molecular orbitals. Electrons in this type of orbital are not concentrated in between the two nuclei of the bonded atoms.

aqueous solutions – Solutions in water.

G

atom – The smallest particle of an element that retains the chemical properties of that element.

atomic mass (atomic weight) – The mass in grams of one mole of an element.

atomic number (Z) – The number of protons in the nucleus of an atom.

Avogadro's constant – The magic mole number, 6.022×10^{23}. The number of carbon-12 atoms in exactly twelve grams of carbon-12.

base – A substance that contains hydroxide ions.

bond energy (BE) – The energy needed to break the bonds between the atoms in one mole of a substance.

bond length – The distance between two bonded atoms.

bond order – Bond order refers to the stability of a covalent molecule, and is equal to one-half the difference between the number of bonding electrons and the number of antibonding electrons.

GLOSSARY

The Super-Charged World of Chemistry Parts 1, 2 & 3

bonding molecular orbital – Concentrates electrons between the nuclei.

Boyle's law – States that the volume of a sample of gas at a given temperature varies inversely with applied pressure.

calorimetry – Measures the heat absorbed or created in chemical reactions.

cation – An ion with a positive charge.

change of state – The movement of matter from one state to another, such as water evaporating (liquid to gas) or ice melting (solid to liquid).

chemical property – Property that describes how a substance reacts or changes into another substance (such as water breaking down into the elements hydrogen and oxygen).

chemistry – The study of the stuff around us and how it changes. The study of how matter and energy interact.

colligative properties – The properties of solutions that are determined by the concentration of solute in a solvent.

G

compound – A substance with two or more kinds of atoms combined in fixed proportions.

covalent bonds – A chemical bond formed between two atoms that share a pair of electrons.

crystal lattice – The repeating pattern formed by the atoms or ions in a crystal.

Dalton's law of partial pressures – States that in a mixture of gases, the total pressure the mixture exerts is the sum of the pressures that each gas would exert if it were alone under the same conditions.

diffusion – The process by which one gas mixes with another.

dilution – Making a solution less concentrated by adding solvent.

dimensional analysis – Problem solving by converting units without altering the values themselves.

dipole – A molecule that has one slightly negative end and one slightly positive end (also known as a polar molecule). Results from an unequal sharing of electrons.

GLOSSARY

The Super-Charged World of Chemistry Parts 1, 2 & 3

dipole-dipole forces – Forces that result from the slight charges on the ends of polar molecules. Dipole-dipole forces cause polar molecules to stick together.

effusion – The process by which a gas escapes from a container through a very small hole.

electrolyte – A solute that produces ions and thus causes the solution to conduct electricity.

electron – A negatively charged particle outside the nucleus of an atom.

electron-deficient compounds – Compounds with less than eight electrons around the central atom.

electronegativity – In a molecule, an atom's ability to pull the electrons of a covalent bond closer to itself.

electron shell – A group of orbitals with the same value of n.

element – A substance that cannot be broken down chemically into a simpler substance.

G

empirical formula – A formula that shows the ratio of atoms in a compound.

endothermic reaction – A reaction in which heat flows into the system.

endpoint – The point at which an acid-base indicator changes color.

enthalpy – The amount of heat absorbed or released in a system at constant pressure.

equivalence point – The point in an acid-base titration when equal amounts of hydrogen ions and hydroxide ions have been mixed.

exothermic reaction – A reaction in which heat flows out of a system.

expanded valence – Occurs when the number of electrons in an atom's valence shell exceeds eight. This can only happen when n, the principle quantum number, is greater than or equal to three.

GLOSSARY

festination – Speeding up, such as when you walk faster and faster down a steep hill.

formal charge – The charge on a molecule or ion when the bonding electrons are equally shared.

formula weight – The combined mass of all the atoms in one mole of an ionic compound.

gas – Matter with no set volume or shape.

gas (beans) – Caused by cans of Pork 'n Beans.

heat capacity – The amount of heat required to raise the temperature of a substance by one Kelvin.

Hess's Law – The total enthalpy change of a reaction equals the sum of the enthalpy changes of the intermediate steps of the reaction.

heterogeneous matter – Matter with different physical properties throughout.

homogeneous matter – Matter with the same physical properties throughout.

hybrid orbital – Combinations or hybridizations of the atomic orbitals that form around the central atom of molecule.

hybrid orbital number – The number of bonds and unbonded electron pairs that form around a molecule's central atom.

hydrogen bonds – Intermolecular bonds that occur between hydrogen atoms (in one molecule) and electronegative atoms such as oxygen or nitrogen (in another molecule or in a different part of the same molecule).

immiscible liquids – Liquids that are insoluble in each other.

intermolecular forces – The forces of attraction between molecules.

ion – A positively or negatively charged atom.

ion-dipole interaction – The force between charges on an ion and a neutral polar molecule (a molecule with a positive and negative end). Occurs when a solid, ionic substance is dissolved in a liquid consisting of polar molecules.

ionic bond – A chemical bond formed by the electrostatic attraction between positive and negative ions.

isotopes – Atoms of an element that have different numbers of neutrons but the same number of protons in their nuclei.

kakidrosis – Body odor.

law of the conservation of mass – The principle that atoms are neither created nor destroyed in chemical reactions.

Lewis structure (electron dot structure) – Diagram of the covalent bonds and electron pairs in a molecule's structure.

limburger – Gouda's stinky cousin.

limiting reagent – The reactant that limits the amount of product formed.

G

London dispersion forces – Intermolecular forces that result from fluctuating, induced dipoles.

lucubration – Studying or intense work late at night, particularly by the light of a lamp.

mass number – The total number of protons plus neutrons in the nucleus of an atom.

mass percentage – The number of grams of solute in 100 grams of solution.

matter – Anything that takes up space and has mass.

miscible liquids – Liquids that can dissolve in each other.

molality (m) – The number of moles of solute per kilogram of solvent.

molar heat capacity – The amount of heat required to raise the temperature of one mole of a substance by one Kelvin.

molar mass – The combined mass of the atoms in one mole of a substance.

molarity (M) – The number of moles of solute per liter of solution.

mole – A unit of measurement that represents 6.022×10^{23} things.

molecular formula – A chemical formula that tells you the exact number of each type of atom in one molecule of the substance.

molecular geometry – The shape formed by the atoms in a molecule.

molecular mass – The combined mass of the group of atoms that make up a single molecule.

molecule – A single unit of a compound (two or more atoms).

mole fraction (X) – The number of moles of a compound divided by the total number of moles in the mixture.

natural abundance – The relative amount of an isotope that occurs in nature.

neutral solution – A solution in which the amount of acid equals the amount of base, so that the mixture is neither a base nor an acid.

neutron – An electrically neutral particle in the nucleus of an atom.

nucleus – The small, dense portion of an atom where the protons and neutrons reside.

octet rule – States that when a compound forms, an atom gains or loses electrons until it has eight electrons in its valence shell.

osmosis – The flow of solvent through a semipermeable membrane. Solvent flows from the more dilute concentration to the more concentrated one.

partial pressure of a gas – The pressure exerted by one of the gases in a mixture.

penelopize – Doing something over again to stall for time.

GLOSSARY

percent composition – The mass ratio of elements that make up a compound, multiplied by 100 to give a percentage.

percent yield – The actual yield divided by the theoretical yield, times 100.

phase change – The transformation of a substance from one state of matter to another.

phenolphthalein – An acid-base indicator that turns pink when a solution reaches the equivalence point.

physical properties – Properties we can measure without changing the identity of the matter.

polar bear – A big, white bear that lives in arctic places.

polar covalent bond – A covalent bond between two atoms that do not share electrons equally.

precision – The reproducibility of a result.

pressure (of a gas) – The force a gas exerts on a unit of area.

G

proton – A positively charged particle found in the nucleus of an atom.

qualtagh – The first person you see after you leave your house.

quantum mechanics – The theory of the behavior of electrons in atoms and molecules.

reaction enthalpy – The heat change associated with a reaction at constant pressure. Equal to the sum of the bond energies of the bonds broken during a reaction, minus the sum of the bond energies of the bonds formed during the reaction. Can be demonstrated using the First Law of Thermodynamics.

reagents – The substances or compounds that take part in a chemical reaction.

resonance structures – Two or more varying (but equally correct) versions of a molecule's Lewis structure. Averaged together for the accepted Lewis structure.

saturated solution – A solution at equilibrium in which the maximum amount of solute has been dissolved; any additional solute will remain undissolved.

Sharky's Machine – The 1970's Burt Reynold's film in which the villain is pushed from the tallest hotel in the world.

significant figures – The numbers on either side of the decimal point that tell how exact a measurement is.

sneezeweed – A perennial yellow flower that causes spewing sickness.

solubility – The maximum amount of a substance that can dissolve in a particular solvent at a particular temperature.

solute – In a solution, the substance(s) present in smaller quantities. When a solution consists of a solid dissolved in a liquid, the solid is always the solute.

solution – A homogeneous mixture of substances.

solvent – In a solution, the substance present in the largest amount. When a solution consists of a solid dissolved in a liquid, the liquid is always the solvent.

specific heat – The amount of heat required to raise the temperature of one gram of a substance by one Kelvin.

standard enthalpy of formation ($\Delta H°_f$) – In a reaction, the enthalpy change that forms a compound from elements in their standard states.

standard state – A substance is in its standard state when it is at $25°C$, one atmosphere of pressure, and in its most stable form.

stoichiometry – Relationships between quantities of material in chemical reactions.

subshell – One or more orbitals that have the same values for n and l.

substance – Homogeneous matter.

surd – A quantity that you can't express in rational numbers.

surroundings – Everything but the system taking part in a reaction.

system – All the substances taking part in a reaction, plus the reaction vessel.

theoretical yield – The amount of product expected from a reaction, based on its balanced equation.

thermochemistry – Measures the heat changes associated with chemical reactions.

titration – A method for determining the concentration of a solution. Titration involves adding an unknown concentration to a known concentration until the mixture reaches the equivalence point.

triple point – Represents the pressure and temperature at which all three phases of a substance are at equilibrium.

ucalegon – Your neighbor whose house is on fire.

unit factor method – A process of converting units without changing values.

valence shell – The outermost shell of an atom.

vapor pressure – In a closed system, the constant force that a vapor exerts above a liquid (once equilibrium has been established).

STUDY SIDEKICK

348

VSEPR model – A theoretical model describing how electron pairs bond. Used to predict molecular geometry.

Zeigarnik effect – The habit of remembering what you haven't done and forgetting what you actually have done.

I've seen the future and I've left it behind.

– Ozzy